FRANCE AT THE CRYSTAL PALACE

FRANCE AT
THE CRYSTAL PALACE

*Bourgeois Taste and Artisan Manufacture
in the Nineteenth Century*

WHITNEY WALTON

UNIVERSITY OF CALIFORNIA PRESS
BERKELEY   LOS ANGELES   OXFORD

University of California Press
Berkeley and Los Angeles, California

University of California Press, Ltd.
Oxford, England

Library of Congress Cataloging-in-Publication Data

Walton, Whitney.
    France at the Crystal Palace : bourgeois taste and artisan
manufacture in the nineteenth century / Whitney Walton.
        p.   cm.
    Includes bibliographical references and index.
    ISBN 0–520–07692–3 (alk. paper)
    1. Consumption (Economics)—France—History—19th
century. 2. Middle classes—France—History—19th century.
3. France—Manufacturers—History—19th century. 4. Industry and state—France—History—19th century.   I. Title.
HC280.C6W35 1992
338.4'767'094409034—dc20   91–41181
                                                                    CIP

Printed in the United States of America
9 8 7 6 5 4 3 2 1

Portions of chapter 2 appeared in *Business History Review* 60
(Winter 1986) under the title " 'To Triumph before Feminine
Taste': Bourgeois Women's Consumption and Hand Methods of Production in Mid-Nineteenth-Century Paris"; copy-
right © 1986 by the President and Fellows of Harvard Col-
lege. A version of chapter 7 appeared in *French History* 3
(1989) under the title "Political Economists and French Industrialization during the French Second Republic (1848–
1852)."

In memory of David

Contents

Illustrations

Acknowledgments

I am very grateful to those institutions that supported the research for this book. A predoctoral fellowship from the American Association of University Women allowed me to spend a year in France doing the research for my dissertation, on which this book is based. An Oakland University Faculty Summer Research Fellowship and a National Endowment for the Humanities Travel to Collections Grant were invaluable in supporting the additional research necessary for revising the dissertation into a monograph.

Writing the dissertation and the book took many years—far more than I care to count—and many persons graciously assisted me at various stages of the process with their time, insights, and suggestions. I wish to thank Edward T. Gargan, my dissertation advisor, for constant encouragement and the freedom to explore a topic that was not fashionable when I was in graduate school. I also thank Domenico Sella for serving on my dissertation committee and for reading and commenting upon portions of the manuscript. I am grateful to Rondo Cameron and Rosalind Williams who were very generous in reading and responding to drafts of articles that were incorporated into the book. I am deeply indebted to friends and scholars who read parts of the manuscript and offered excellent comments and much-needed encouragement; my sincere thanks to Ellen Furlough, Julia Green, Daryl Hafter, Scott Haine, Miriam Levin, Tessie Liu, John Merriman, and Leslie Page Moch. Special thanks to Christopher Johnson for reading so much, for always offering pertinent suggestions, and for providing in his own work a model of fine scholarship. I am no less grateful to the many other friends who talked with me about consumption and production in nineteenth-century France; they deepened my understanding of a complex topic and enhanced my confidence in the whole enterprise. I owe particular thanks to Anne Chapple, who obtained the illustration for the dust jacket.

I very much appreciate my family's support during the many years of writing and revising; their belief in me and in the eventual completion of the book was invaluable. My husband Thomas Brush also helped me in more ways than I can list here; his good humor, insightful remarks, steady encouragement, and computer competence are only some of the many qualities that carried me through the long process.

Introduction

French Taste and Manufacturing at the Crystal Palace Exhibition

In London on May Day 1851 the sun burst through the clouds just as Queen Victoria approached Hyde Park in the royal carriage. The glass panes of the great exhibition hall gleamed in the midday light, colored banners flew from the top of the structure's iron girders, and Victoria recorded in a letter that "the sight of the Crystal Palace was indescribably glorious, really like fairyland."[1] She and her entourage of royal children, ladies-in-waiting, and high officials entered the building and made their way through cheering crowds to a velvet-covered dais located at one end of the wide central transept and facing the quietly splashing crystal fountain across the assembled throng. After cannonades, a rousing chorus of "God Save the Queen," and a prosaic speech by the prince consort on the purpose and history of the event, the queen ceremoniously opened the Great Exhibition of the Works of Industry of All Nations.

The Crystal Palace exhibition of 1851, the first in a long series of international industrial fairs, was unprecedented in size and scope. Measuring 1,851 feet long and 456 feet wide, the glass and iron structure that housed the exhibition was four times the length and twice the width of St. Paul's Cathedral, and a brilliant product of modern industrial methods and technology (fig. 1). Joseph Paxton's innovative design permitted 1,800 workers to assemble the standard-ized panes and girders in a matter of months, and the almost over-night appearance of the grand, seemingly fragile exhibition hall led a

1. Quoted in C. R. Fay, *Palace of Industry, 1851* (Cambridge: Cambridge University Press, 1951), 25.

writer for *Punch* to christen the wondrous edifice with the evocative and enduring sobriquet of Crystal Palace (fig. 2).[2]

No less impressive than its exterior was the sumptuous array of capital and consumer goods which filled the inside of the Crystal Palace from 1 May through 15 October. Approximately 15,000 exhibitors from all over the world proffered raw materials, machines, and manufactured products to the admiration of an average of 43,000 visitors daily. While exhibits from Britain and her colonies occupied fully one-half of the floor space in the Crystal Palace—both the ground floor and the galleries of the western or British nave—the other half was divided among participants from France, Austria, Switzerland, Belgium, the Zollverein, Denmark, Russia, Italy, the Netherlands, Spain, Portugal, Turkey, Egypt, Greece, Persia, Brazil, China, and the United States (fig. 3). Differences in the exhibits from industrialized and less industrialized countries were readily apparent. Russia, Canada, and the United States, for example, were particularly strong in displays of raw materials—gold, platinum, iron, timber, furs, skins, corn, rye, oats, barley, cotton, tobacco, and so forth. Not surprisingly, the British dominated in the general category of machines, though notable exceptions included a reaper by Cyrus McCormick of the United States and an envelope-making machine from France.[3]

More plentiful than raw materials and machines, however, and extremely popular among spectators, were the various and often splendid manufactured goods produced by hand and by machine. Here, the gorgeous products of painstaking manual labor from India and China were as much admired as, and often more beautiful than, the machine-made articles from industrially advanced countries like England. In any case, all manner of visitors to the Crystal Palace enjoyed viewing the extraordinary examples of everyday items of furnishing and clothing. Fine silks, colorful cottons, sturdy woolens, delicate lace, rich embroidery, and brilliant shawls attracted the attention of women and men, royalty and commoners. Sculpted cab-

2. Fay, *Palace*, 15; John Tallis, *Tallis's History and Description of the Crystal Palace* (London and New York: John Tallis and Co., [1852]) 1:9–13.

3. Fay, *Palace;* Tallis, *Tallis's History;* Yvonne Ffrench, *The Great Exhibition: 1851* (London: Harvill Press, 1950); E. Eldon Hall, *A Condensed History of the Origination, Rise, Progress and Completion of the "Great Exhibition of the Industry of All Nations"* (Clinton Hall, N.Y.: Redfield, 1852); Christopher Hobhouse, *1851 and the Crystal Palace* (New York: E. P. Dutton and Co., 1937).

inetry, lacquered and papier-mâché furniture, interior and exterior ironwork, jewelry of precious metals and gems, bronze art objects, glazed and painted porcelains, carved boxes, intricate fans, and a host of other consumer goods also pleased the curious hordes (fig. 4). The Crystal Palace exhibition, like the international shows that followed it, was a showcase of industrial capacity to satisfy a range of consumer tastes from the fairly simple to the most luxurious.

However, historians have neither examined the Crystal Palace exhibition from a consumer-oriented perspective nor analyzed the relationship between consumer demand and production methods that the exhibition unveiled.[4] Instead, interpretations of the exhibition have emphasized its significance in heralding the triumph of mass production methods in England, implying that this would be the model for subsequent developments in other industrializing nations of Europe.[5] Like many economic and social histories, these interpretations focus almost exclusively on production and assume a framework of inexorable progress toward greater efficiency in manufacturing, at the expense of small-scale, hand methods of production and an artisan working class. According to this view, England led the way toward mass mechanized, concentrated, large-scale production under the direction of enterprising capitalists, and other countries would sooner or later have to follow this pattern in order to compete successfully in world markets.[6] Recently, however, scholars have

4. Rosalind H. Williams has admirably analyzed the rise of mass consumption in later French exhibitions of the 1880s and 1890s in her book *Dream Worlds: Mass Consumption in Late Nineteenth-Century France* (Berkeley: University of California Press, 1982), but she does not address the relationship of consumption to production, nor the gender relations involved in consumption. An innovative analysis of the Paris exhibitions of 1889 and 1900 that includes consumer taste and women as consumers is Debora L. Silverman, *Art Nouveau in Fin-de-Siècle France: Politics, Psychology, and Style* (Berkeley: University of California Press, 1989).

5. David S. Landes, *The Unbound Prometheus* (London: Cambridge University Press, 1969); Arthur Louis Dunham, *The Industrial Revolution in France, 1815–1848* (New York: Exposition Press, 1955). Cultural and social interpretations of the Crystal Palace exhibition include Jean Duvignaud, *Fêtes et civilisations* (Geneva: Weber, 1973); Richard D. Altick, *The Shows of London* (Cambridge, Mass.: Belknap Press, 1978); Werner Plum, *Les Expositions universelles au 19e siècle, spectacles du changement socio-culturel*, trans. Pierre Gallissaires (Bonn and Godesberg: Friedrich-Ebert-Stiftung, 1977).

6. William Otto Henderson, *Britain and Industrial Europe, 1750–1870: Studies in British Influence on the Industrial Revolution in Western Europe*, 3d ed. (Leicester: Leicester University Press, 1972).

disputed this model of industrial capitalist development and have shown that both in England and in other parts of Western Europe alternative structures of manufacturing—small-scale, dispersed, only partially mechanized, and capable of producing short runs of goods that varied in design and style—in some instances proved highly successful in the nineteenth and twentieth centuries.[7] If this is indeed true—and the evidence marshaled is convincing—then what did the Crystal Palace exhibition really signify in the history of industrializing society? In light of this new research on manufacturing methods and organization, it is appropriate to reexamine the first international exhibition of industry with greater attention to the varieties of production processes represented there.

But to focus only on production at the exhibition would be, again, a mistake. The new research on alternatives to mass production mentions, but too rarely pursues, the fact that certain methods of manufacturing succeeded in part because they satisfied certain demands. Thus the full story of why parcelized, hand, small-scale, or specialized production processes were successful requires an understanding of consumer tastes and motivations. This type of inquiry is particularly relevant to the case of France, where parcelized structures of manufacturing (as opposed to mass production methods) predominated at the time of the exhibition and for several decades thereafter.[8] The Crystal Palace exhibition, by allowing for the first time a comparison of French and British industry, could have been a major

7. Charles F. Sabel and Jonathan Zeitlin, "Historical Alternatives to Mass Production," *Past and Present* 108 (1985): 133–76; Maxine Berg, Pat Hudson, and Michael Sonenscher, *Manufacture in Town and Country before the Factory* (New York: Cambridge University Press, 1983); Ronald Aminzade, "Reinterpreting Capitalist Industrialization: A Study of Nineteenth-Century France," in *Work in France: Representations, Meaning, Organization, and Practice,* ed. Steven Laurence Kaplan and Cynthia J. Koepp (Ithaca, N.Y.: Cornell University Press, 1986), 393–417; Alain Faure, "Petit Atelier et modernisme économique: La Production en miettes au XIXe siècle," *Histoire, économie et société* 4 (1986): 531–57; Raphael Samuel, "Workshop of the World: Steam Power and Hand Technology in Mid-Victorian Britain," *History Workshop* 3 (Spring 1977): 6–72.

8. T. J. Markovitch calculates that nearly 60 percent of all French industrial production as late as 1860 was hand or handicraft in nature, occurring in homes or small workshops where artisan manufacturers worked alongside fewer than ten employees. T. J. Markovitch, "Le Revenu industriel et artisanal sous la Monarchie de Juillet et le Second Empire," *Économies et sociétés,* série AF-8 (April 1967): 85–86.

turning point for French manufacturing, leading it in the direction of mass mechanized production after the model of certain industries in England, like textiles. But the majority of French commentators on the exhibition viewed it as a vindication of small-scale manufacturing in France, with its emphasis upon the stylishness and quality of products rather than on their low prices or large quantities. In addition, jury members and reporters from France frequently credited the success of French industry at the Crystal Palace to the good taste of French consumers, and the issue of taste in consumption and production subsequently became an important part of the discourse in France over national economic policy and industrial development.[9] This examination of French documentation of the exhibition, along with other contemporary sources, recasts the conceptualization of French industrial development. It argues that the desires and concerns of bourgeois consumers in many cases promoted parcelized, hand methods of manufacturing in France, and that the successful performance of French industry at the exhibition led to policy proposals intended to support this pattern of industrial development.

Consumer demand, of course, was not the only reason for the persistence of hand manufacturing in nineteenth-century France. Economic historians have identified several other factors contributing to the slow pace of mechanization and concentration in French industry, including insufficient and inaccessible raw materials (especially coal), inadequate financial and credit institutions, a propertied peasantry not solely dependent on industrial wage labor, deficient communication and transportation networks, and a cautious, secretive entrepreneurial class.[10] Implied or explicit in all of these explanations is the assumption of the English model or norm, in

9. The French were not alone in their evaluation of the importance of consumer and producer taste in manufacturing. The English conceived of the exhibition as a test of just how superior French manufacturers were in the matter of taste, and in what ways the British could encourage better taste in their own manufacturing. Tobin Andrews Sparling, *The Great Exhibition: A Question of Taste* (New Haven: Yale Center for British Art, 1982), ix, 1. See also Toshio Kusamitsu, "British Industrialization and Design before the Great Exhibition," *Textile History* 12 (1981): 77–95.

10. Roger Price, *The Economic Modernisation of France* (New York: Wiley, 1975); Rondo E. Cameron, *France and the Economic Development of Europe, 1800–1913* (Princeton: Princeton University Press, 1961); Landes, *Unbound Prometheus*.

comparison with which France was lacking or retarded.[11] More recently scholars have justifiably abandoned the framework of French retardation, though they continue to compare economic performance in Britain and France in the nineteenth century.[12] Shifting from French deficiencies, another perspective suggests that differences in labor supply and conditions informed varied patterns of industrialization on a regional even more than a national basis. According to this theory, entrepreneurial strategies for manufacturing depended in large measure not only upon the availability and cost of labor, but also upon the family economies of workers, the structures of female labor, and the ways that workers perceived and pursued their best interests.[13] While conditions surrounding production were obviously critical in influencing French industrialization, consumer demand also affected management decisions and labor relations.

Scholars have noted that, given the preponderance of subsistence or wage-earning peasantry in the French population of the nineteenth century, the consumers of manufactured goods were primarily middle class. Bourgeois consumers, they indicate, were fairly wealthy and cultivated; they demanded fashionable, high-quality goods that factory manufacturing could not always produce. Hence manufacturers had no incentive to adopt mass production processes.[14] Though demographically this consuming class was very small, its importance as a market for French manufactured goods, in particular the luxury products made in Paris, was very great.

---

11. Alexander Gerschenkron, *Economic Backwardness in Historical Perspective* (Cambridge, Mass.: Belknap Press, 1962); Charles P. Kindleberger, *Economic Growth in France and Britain, 1851–1950* (Cambridge: Harvard University Press, 1964).

12. Richard Roehl, "French Industrialization: A Reconsideration," *Explorations in Economic History* 13 (1976): 233–81.

13. Alain Cottereau, "The Distinctiveness of Working-Class Cultures in France, 1848–1900," in *Working-Class Formation: Nineteenth-Century Patterns in Western Europe and the United States,* ed. Ira Katznelson and Aristide R. Zolberg (Princeton: Princeton University Press, 1986), 111–54.

14. Patrick O'Brien and Caglar Keyder, *Economic Growth in Britain and France, 1780–1914: Two Paths to the Twentieth Century* (London: George Allen and Unwin, 1978); Maurice Lévy-Leboyer, "Les Processus d'industrialisation: Le Cas de l'Angleterre et de la France," *Revue historique* 239 (1968): 281–98; Rondo E. Cameron, "Economic Growth and Stagnation in France, 1815–1914," in *The Experience of Economic Growth: Case Studies in Economic History,* ed. Barry E. Supple (New York: Random House, 1963), 328–39.

The results of two surveys of Parisian industry, conducted in 1847–48 and 1860 by the Paris Chamber of Commerce, indicate that prior to 1860 Parisian manufacturers supplied primarily Parisian consumers, and that the local market never ceased to be more significant than foreign sales. The 1847–48 survey noted: "What is immediately striking . . . is the importance that local consumption must have, compared to the value of goods produced in Paris and sent either to the rest of the country or abroad."[15] By 1860, according to the second survey, the proportion of Parisian goods exported had increased, though domestic consumption continued to absorb at least three-quarters of most types of manufactured products. Fourteen percent of the furnishings produced and 17 percent of the clothing were exported. The classification that exported the most among Parisian manufactures, at 26 percent, was *articles de Paris,* which included fans, little decorative boxes, buttons, artificial flowers, fine combs, and other small luxury and decorative items.[16] In her study of Paris under the Second Empire, Jeanne Gaillard emphasized how important exports, particularly of luxury goods, were to Parisian industrial growth after 1860 — they comprised an estimated 16 to 17 percent of total French exports in the opening years of that decade — but she also notes the novelty of this phenomenon.[17] Renowned internationally for their artistic merit, elegance, and high quality, consumer goods made in Paris might never have acquired this reputation abroad had not local and national consumers demanded such characteristics prior to and well beyond 1860.

What made French and especially Parisian consumers so exigent about the beauty and quality of home furnishings and clothes? What were the different production methods and structures that bourgeois demand promoted in French consumer goods industries? And how did government policy makers interpret and affect this relationship? Answers to these questions from French documentation of the Crystal Palace exhibition establish a direct link between bourgeois taste and hand manufacturing in nineteenth-century France. Moreover,

15. Chambre de Commerce et d'Industrie, *Statistique de l'industrie à Paris, 1847–48* (Paris, 1851), 38.

16. Chambre de Commerce et d'Industrie, *Statistique de l'industrie à Paris de l'enquête faite par la Chambre de Commerce pour l'année 1860* (Paris: La Chambre, 1864), xlv–xlvi.

17. Jeanne Gaillard, *Paris, la ville, 1852–1870* (Paris: H. Champion, 1977), 380–85.

they indicate that women, as well as men, of the bourgeoisie influenced production through consumer demand, and that possessing good taste was an integral component of bourgeois status.

Studies of the French bourgeoisie in the nineteenth century, based on a notion of the bourgeois as capitalists or owners of the means of production, have often focused on the rise of male manufacturers and politicians to positions of economic and political prominence.[18] Other studies, using a cultural definition of the bourgeoisie as a class sharing certain values with respect to individual achievement and family unity and enjoying a comfortable standard of living from earned or invested income, have presented a more detailed picture of the private lives, personal fortunes, and public activities of bourgeois men and women.[19] But none have addressed the issue of consumption, an essential facet of bourgeois identity and existence.[20]

This study proposes to define the bourgeoisie in terms of its consumer tastes and practices, and to demonstrate that these values and habits upheld a social order that the bourgeoisie aspired to dominate. Examining taste will be a means of illuminating the social and gender relations inherent in the identity of a class, and of rean-

18. Guy P. Palmade, *French Capitalism in the Nineteenth Century,* trans. Gramme M. Holmes (Newton Abbot, England: David and Charles, 1972); Claude Fohlen, *L'Industrie textile au temps du Second Empire* (Paris: Plon, 1956); Jean L'homme, *La Grande Bourgeoisie au pouvoir, 1830–1880* (Paris: Presses universitaires de France, 1960).

19. Michelle Perrot et al., *De la Révolution à la Grande Guerre,* vol. 4 of *Histoire de la vie privée,* ed. Philippe Ariès and Georges Duby (Paris: Éditions Seuil, 1987); Adeline Daumard, *Les Bourgeois et la bourgeoisie en France depuis 1815* (Paris: Aubier, 1987); Peter Gay, *The Bourgeois Experience from Victoria to Freud,* vol. 1, *Education of the Senses* (New York: Oxford University Press, 1984); Bonnie G. Smith, *Ladies of the Leisure Class: The Bourgeoises of Northern France in the Nineteenth Century* (Princeton: Princeton University Press, 1981); Adeline Daumard, *La Bourgeoisie parisienne de 1815 à 1848* (Paris: SEVPEN, 1963).

20. One exception is Theodore Zeldin, *Taste and Corruption,* vol. 4 of *France, 1848–1945* (Oxford: Oxford University Press, 1980), though his analysis of bourgeois consumption is sketchy at best. See Whitney Walton, "'To Triumph before Feminine Taste': Bourgeois Women's Consumption and Hand Methods of Production in Mid-Nineteenth-Century Paris," *Business History Review* 60 (Winter 1986): 541–63. An important work on bourgeois consumption in the United States is Susan Porter Benson, *Counter Cultures: Saleswomen, Managers, and Customers in American Department Stores, 1890–1940* (Urbana: University of Illinois Press, 1986). For a fascinating explanation of the origins of the English Industrial Revolution in the eighteenth-century demand for cotton calicoes, see Chandra Mukerji, *From Graven Images: Patterns of Modern Materialism* (New York: Columbia University Press, 1983), esp. chaps. 5–6.

alyzing the role of the bourgeoisie in French industrial development. To be sure there were great economic and social disparities within the bourgeoisie, from families possessing substantial wealth, superior education, and political influence, to those having a modest income, limited education, and little power. The premise is, however, that these groups were united by an appreciation for the significance of material possessions as symbols of bourgeois status.

As some social scientists have shown, the goods people acquire for personal adornment and home furnishing are inextricably linked with social and political relations.[21] By analyzing the tastes and possessions of French consumers as revealed in exhibition documentation, it is possible to understand more about the rise of the bourgeoisie to ruling-class status, the efforts of this class to distinguish itself from the aristocracy and the working class, the power struggles between men and women in a society with fairly rigid but often contradictory gender roles, and the interaction of bourgeois demand and different methods of production. This understanding of the social and gender relations behind French commentators' preoccupation with good taste and social status helps explain why certain methods of manufacturing were more successful in some French consumer goods industries than in others. The Crystal Palace exhibition presented objects made from new materials and by new processes, but inventories of household possessions indicate that French bourgeois at midcentury were selective in their adoption of these innovations. Only after satisfying their craving for status and the elegance of handmade goods did consumers appropriate the newer objects that met other criteria such as comfort, practicality, and low cost. The exhibition showed that in general, skilled handicraft and dispersed manual methods of manufacturing were more suitable to meeting bourgeois demand for artistic, stylish articles of furnishing and clothing than mass, mechanized production. However, certain mechanical

21. Arjun Appadurai, ed., *The Social Life of Things: Commodities in Cultural Perspective* (New York: Cambridge University Press, 1986); Daniel Miller, *Material Culture and Mass Consumption* (Oxford and New York: Basil Blackwell, 1987); Pierre Bourdieu, *La Distinction: Critique social du jugement* (Paris: Éditions de minuit, 1979). Valuable earlier works on consumption include Georg Simmel, *The Philosophy of Money* (1904), trans. Tom Bottomore and David Frisby (London and Boston: Routledge and Kegan Paul, 1978); Thorstein Veblen, *The Theory of the Leisure Class* (New York: Macmillan Co., 1899). See chapter 1 for further discussion of theoretical works on consumption.

or otherwise new processes capable of accommodating bourgeois notions of taste met with early favor and success. Bourgeois consumers, then, influenced the production processes employed in the manufacture of different household goods, though new technology and new work organizations also contributed to changes in the material life of the bourgeoisie. This two-way relationship will be addressed by linking consumer values with the methods of production explained in the exhibition documentation.

The exhibition was also a focal point for discussion of the future of French industrial and economic policies. As a result of France's stellar performance at the Crystal Palace most academics, industrialists, and statesmen were satisfied that French manufacturing, with its many small-scale and hand industries as well as some modern factories, was economically viable and competitive at that time. In fact their proposals for French economic policy were explicitly intended to preserve France's reputation for good taste in manufacturing while allowing for some change toward mechanized, mass production methods. The books and articles that French policy makers wrote in connection with the Crystal Palace indicate that quality and good taste in production—most often found in small-scale, hand manufacturing—was a clearly articulated goal of many men in a position to influence government policy regarding industrial development. Disagreement over the means to achieve this end, however, hindered the implementation of any major government intervention in economic direction. But it is reasonable to suppose that France's successful participation in the Crystal Palace exhibition ultimately contributed to the Anglo-French Agreement of 1860—the culmination of a long campaign for free trade in France and one of the most significant events in French economic policy of the nineteenth century.

The Crystal Palace exhibition of 1851 encapsulated in a single event a range of social tensions, economic changes, and political struggles occurring in mid-nineteenth-century France.[22] It was also a

22. The analysis of single events or texts as a means of deciphering larger social, cultural, and political issues is a method fruitfully employed in the following works: Emmanual Le Roy Ladurie, *Carnival in Romans,* trans. Mary Feeney (New York: George Braziller, 1976); Robert Darnton, *The Great Cat Massacre and Other Episodes in French Cultural History* (New York: Basic Books, 1985); Joan W. Scott, "Statistical Representations of Work: The Politics of the Chamber of Commerce's *Statistique de l'industrie à Paris, 1847–48,*" in *Work in France,* ed. Kaplan and Koepp, 335–63.

matter of considerable concern to those French men and women of the time who saw in the exhibition an opportunity for France to assert a leading role in a global economy. The French took an avid interest in the first international exhibition of industry because, for one thing, the idea and practice of industrial exhibitions originated in France under the First Republic (1792–1804). Since that time the French government had sponsored exhibitions of national industry approximately every five years to encourage technological development and economic growth in the country.[23] While planning for the exhibition of 1849, the minister of agriculture and commerce, Louis-Joseph Buffet, proposed that it should be open for the first time to foreign participants; but others cited the recent revolution of 1848 and the policy of economic protectionism in France as grounds for rejecting internationalism at that time.[24] A visitor to the 1849 exhibition in Paris, the British manufacturer Henry Cole, also conceived of the notion of an international exhibition of industry. With the support of Prince Albert and other British manufacturers he was successful in launching the project in London in 1851.[25]

Lingering doubts about the benefits of foreign competition caused French industrialists to drag their feet in responding to the government's request for submissions of exhibits to the London exhibition.[26] However, by the time of its opening, educated French men and women regarded the exhibition as a test of France's renowned manufacturing prowess against the emerging industrial might of Great Britain.[27] Statesmen in particular automatically assumed that

23. [Adolphe] Blanqui, "Expositions des produits de l'industrie," in *Dictionnaire de l'économie politique,* ed. C. Coquelin and Guillaumin (Paris: Guillaumin et Cie, 1852–53) 1:746–49.

24. Georges Gerault, *Les Expositions universelles envisagées au point de vue de leurs résultats économiques* (Paris: Larose, 1902), 31, 34; Utz Haltern, *Die londoner Weltausstellung von 1851* (Münster: Verlag Aschendorff, 1971), 129–30.

25. Hobhouse, *1851,* 7–8.

26. Ministère de l'Intérieur de l'Agriculture et du Commerce, *Annales du commerce extérieur: Faits commerciaux,* no. 20, *Exposition universelle de Londres en 1851* (Paris: Imprimerie impériale, 1853), 7.

27. As early as February 1851 a women's periodical asserted: "Each day our French industry creates new marvels for the London Exhibition—each day new masterpieces leave [for England] to represent honorably our taste and our skill." *Petit Courrier des dames,* 22 February 1851. The daily *Le National,* initially skeptical about the international exhibition, wrote in June 1851: "France and the Republic have triumphed together in sending these rich products to the industrial congress." *Le National,* 10 June 1851, 1.

their country's manufacturers would rise to the challenge. Added urgency for France to enter the lists came from the fact of the economic depression and social upheaval from 1846 through 1848; French officials and many manufacturers believed it important to show that peace and prosperity reigned once again in France and that the country was ready and able to do business with the rest of the world.

Upon receipt of Queen Victoria's January 1850 announcement of the exhibition project, the minister of agriculture and commerce, Jean-Baptiste Dumas, wrote to the president of the Republic: "There is no question but that French industry . . . will uphold the renown it has acquired for taste and skill; no question but that our manufacturers will hasten to send those of their products they judge worthy to compete in this tournament and uphold the reputation of France."[28] With the president's approval Dumas set in motion the committees and procedures to solicit, judge, and transport entries of French industrial machines and products to the exhibition.[29] In January 1851 the National Assembly voted an allotment of 638,000 francs to cover the expenses of participation in the exhibition, and a total of 3,518 packages comprising the contributions of 1,760 French exhibitors went to London barely in time for the 1 May opening.[30]

Accompanying the French manufacturers and their goods to the Crystal Palace was a group of fifty-six men appointed by the minister of agriculture and commerce to represent France on the international jury. Composed of twenty-two academics and civil servants, twenty-four industrialists, and ten practitioners of liberal professions, this body spent several weeks at the exhibition examining the products in its members' fields of expertise and meeting with jury members of other nationalities to select the winners of prize medals and honorable mentions on the basis of the quality, innovation, and value of the articles under consideration. The French jury members' tasks also included writing extensive reports on the exhibits and evaluating French products and industry in comparison with those of other countries. After many years' delay the imperial press finally published

28. Ministère de l'Intérieur, *Annales,* 37.
29. Ibid., 6, 7.
30. Ibid., 11, 17.

the multivolume collection of reports which will serve as a basic source for this analysis of French consumption and manufacturing.[31]

The official exhibition reports contain general reflections on industrial development, descriptions and analyses of industry in different parts of the world, and detailed information on all types of industry in France. Each reporter for a specific French trade was an expert in the field—either a successful entrepreneur or an accomplished academic—and in addition to describing the products from France displayed in the Crystal Palace he often narrated the history of the industry, explained the process of production, compared method and quality with foreign analogues, and speculated about future developments in that area. Several writers used data from recent national and municipal investigations of industry and of the working class, rendering the 1851 exhibition reports a composite of current information on many industries and related technological, political, social, and cultural issues.[32] Coverage of the exhibition in the popular press was also extensive, providing added insights into the tastes and concerns of the French reading public regarding the finest consumer products of French manufacture and their performance in the international arena.

More difficult to assess, and of vital importance to the study of bourgeois consumption in nineteenth-century France, is precisely what bourgeois men and women actually bought for themselves, their families, and their homes. Autobiographical sources—diaries, letters, memoirs—that include accounts of personal and family purchases are rare; a systematic search through published works and archival records for information on private consumption must be the basis for a separate book devoted solely to the consumer habits of the bourgeoisie. Since this study seeks to illuminate the connection between bourgeois demand and industrial development, it has relied primarily on prescriptive literature regarding bourgeois consumption, supported by examples from fiction and autobiographies, and on a sample from the notarial records of household possessions located in the National Archives in Paris (see chapter 3). These

31. Ibid., 40–42, 14–15.
32. Commission française sur l'Industrie des Nations, *Exposition universelle de 1851: Travaux de la Commission française sur l'Industrie des Nations,* 8 vols. (Paris: Imprimerie impériale, 1853–56).

sources serve to amplify the understanding of consumer standards and to confirm the selectivity of consumer practices as revealed in the various reports of the exhibition.

Although the exhibition documentation on French industry is copious, certain practical constraints circumscribed this representation. First, the sample of French industrial products in the Crystal Palace was predominantly from the Paris region; of 1,629 continental exhibitors, 990 came from the department of the Seine, and the majority of these manufacturers were involved in some form of dispersed hand or handicraft industry.[33] For practical and geographical reasons, it was relatively easy for Parisian manufacturers of consumer goods to send products to London via Paris; provincial producers at a distance from the capital had to contend with additional transportation costs for their sometimes bulky and heavy capital goods. In addition, many provincial factory owners refused to exhibit in London, fearing that their textiles, metals, and machinery would compare unfavorably in price with British goods. These manufacturers reasoned that if their clients learned they could obtain cheaper products from Britain, the pressure to abolish the existing prohibitions and tariffs on importations would mount, and certain French industrialists would lose their monopoly over the French market. Moreover, mechanically made articles were generally less remarkable than handmade goods, and the jury which selected exhibits favored finer and more exotic products to represent French industry at the Crystal Palace. Given these limitations and biases, the displays from France might still be considered typical of French manufactured products in 1851. In terms of the value of manufactured goods and the number of people involved, small-scale hand production was the most important type of manufacturing in France at this time.

The exhibition itself fulfilled, even surpassed, the expectations of French men and women. A writer for the feminine press urged readers to spend their summer holidays in England, visiting the exhibition that was "more marvelous than the crystal palaces of fairy tales" and that housed "the industrial masterpieces of all the nations of the earth."[34] Reporting for the popular weekly *L'Illustration*, Émile

33. Ministère de l'Intérieur, *Annales*, 17, 20.
34. *Journal des mères et enfants*, 1 August 1851, 79.

Berès confessed: "When I am no longer completely dazzled by all these splendors and all this tumult, how will I be able even to try to communicate to you readers the prodigiousness of this human crowd, aristocratic and shabby, undoubtedly the largest that a building has ever enclosed?"[35] Even Gustave Flaubert, the bourgeois who execrated whatever the bourgeoisie did or liked, called the exhibition "a very beautiful thing," despite the fact that everyone else thought so, too.[36] Praise for the good taste and superior quality of French products at the exhibition was ubiquitous, both in France and elsewhere. Official reports and popular accounts of the exhibition congratulated French producers and consumers on their triumph, especially over the much less tasteful British goods.

Part 1 of this book examines consumption in nineteenth-century France. The first chapter, on the construction of the bourgeoisie as a class, examines the comments on the exhibition displays, revealing the tenets of bourgeois taste in an adherence to traditional styles associated with aristocratic periods in French history and tempered by a concern for domestic comfort. These principles distinguished bourgeois consumers from aristocrats, whose consumption practices emphasized luxury, and from workers, who were limited to the satisfaction of basic necessities. The exhibition clarified and exalted this bourgeois standard of good taste by suggesting alternative taste types or demand criteria, including those of North and South Americans, the British, and the French working class. Critics condemned products intended for these markets as tasteless, but in fact this opinion covered hostility (often thinly disguised) toward those groups, especially French workers, for the challenge they presented to bourgeois domination. Style and quality were more than matters of taste; they represented an entire social, political, and cultural order. This explains why bourgeois commentators were so obsessed with the good taste of French consumers and producers that the exhibition lauded. The exhibition both buttressed the bourgeois notion of good taste and showed the dangers of free-market capitalism in dethroning that standard—and by implication the class it served—by catering to different markets with different tastes. The bourgeoisie confronted a dilemma: it wanted to preserve the exclusive trappings of ruling-class

35. *L'Illustration*, 10 May 1851, 290.
36. Jean Seznec, ed., *Flaubert à l'Exposition de 1851* (London: Oxford at the Clarendon Press, 1951), 23.

status in traditional styles; but in order to obtain the desired goods, it needed the profits from expanded production and trade with different markets.

The exhibition revealed another source of tension within the bourgeoisie and with regard to consumption—gender conflicts. Bourgeois men and women generally agreed that females were inherently more tasteful than men, and that home furnishing and decoration should therefore be confided to their care. This practice was consistent with the bourgeois ideal of domesticity, but it also presented contradictions. Consumption could be, and often was, a source of power for women over men. Moreover, the exhibition itself uncovered the fallacy of feminine consumption for solely domestic or private purposes; it showed that shopping for home and family was also public and required women to enter the marketplace, presumably a space that men dominated. Chapter 2, then, asserts that through consumption bourgeois women influenced manufacturing as much as, if not more than, the men who were exercising their ambitions and desires within patriarchal society. The ascribed and actual feminine role as consumer, while apparently a complementary balance to masculine involvement in production, trade, and politics, was also a source of tension in gender relations.

The third chapter, dealing with consumption, analyzes inventories of bourgeois household possessions to determine which of the industrial innovations and advances demonstrated at the Crystal Palace were actually adopted by bourgeois consumers, and examines some of the reasons behind this selectivity. It tests the conclusions drawn about bourgeois taste in chapters 1 and 2 against the goods people bought for themselves and their homes. Bourgeois consumers placed a high priority on the status symbols of their class and social position, which usually entailed the purchase of certain fairly expensive items of furniture and furnishing. The lesser priorities of low cost and novelty came into play either after the status desires had been fulfilled or where the new materials or products were compatible with the bourgeois conception of good taste.

How did bourgeois demand for consumer goods affect production in France? Part 2 answers this question by examining manufacturing processes in several different industries represented at the exhibition and evaluating their suitability to meet bourgeois consumer standards. Chapter 4 deals with wallpaper making, cabinetmaking, and the fabrication of *articles de Paris;* in these industries various shop and dispersed organizations of hand production prevailed in mid-nineteenth-century France because they satisfied bourgeois demand for stylish, artistic, and fairly high quality items at affordable prices and in fairly large quantities. The market for such items was certainly larger in the nineteenth century than under the Old Regime, with its restrictions on social mobility and consumption, but it was hardly a mass market for unlimited amounts of standardized goods. Official reports on the exhibition explain clearly how hand presses in wallpaper factories, hand construction and sculpting as well as increased division of labor in cabinetmaking, and divided, putting-out hand manufacture along with workshop finishing in *articles de Paris* were the most appropriate techniques for meeting bourgeois demand at midcentury.

By contrast chapter 5 addresses developments in art industries—goldsmithing, silversmithing and bronze making—where mass or mechanized production was more successful than hand manufacturing in satisfying the bourgeois' desires for prestige items they could afford. Electrolytic metal plating in silversmithing and the Collas technique of reproduction in bronze making allowed manufacturers to increase drastically their output and to lower the prices of silver-plated tableware and of bronze statues, respectively. These products, though mass produced, retained the characteristics of luxury and art that bourgeois consumers craved. In cases where modern technology and mass production methods resulted in affordable and artistic goods, bourgeois consumers happily encouraged them.

Part 3 deals with policy proposals for French industrial development that either resulted from or were influenced by the Crystal Palace exhibition. Chapter 6 analyzes the proposals by aristocrats, artisans, and artists to maintain high standards of art and taste in French consumption and production. These men, active participants in the exhibition, concluded from it that French success in worldwide industrial competition depended heavily on the good taste of French consumers and the artistic ability of French producers. They feared that foreign governments, notably that of Great Britain, having witnessed the popularity of tasteful goods from France at the exhibition, would invest in the means necessary to improve the taste and quality of manufactured, including machine-made, products to the detriment of France's international monopoly on good taste. Among the proposals to the French government for combating this threat were exhibitions of industrial art, design schools, public museums, education reforms, and state manufactures. By far the most ambitious suggestions for government intervention to maintain popular taste came from the art critic and exhibition jury member, Count Léon de Laborde. Laborde's plans for the transformation of art and taste in France by using the power of the modern nation-state to impose a traditional aesthetic upon the population are the focus of chapter 6. Referring to the Crystal Palace exhibition and to history as evidence, Laborde argued that only a centralized and interventionist state could "purify" consumer tastes, and so uphold French industrial competitiveness.

Chapter 7 analyzes an opposing view—that of the political economists who responded to the exhibition with the assertion that free trade, not government intervention, would guarantee art and taste in French manufacturing. For decades political economists had been ardent campaigners for free trade in France, believing that it was the best policy for industrial development and economic prosperity. For these bourgeois men, who rose to political prominence partly through the advantages of caste but especially through hard work and personal achievement, free trade was consistent with their bedrock values of individualism, meritocracy, and reformism. In the exhibition they saw a strong support for their cause, but the French performance there also altered their rhetoric concerning free trade, to include explicitly the continuation of quality production—specialized industrialization—in France. Good taste in manufacturing, then, became part of the political discourse of political economy to try to unite bourgeois consumers and working-class producers in one harmonious, capitalist society.

Good taste in consumption and production was a recurring theme in French responses to the Crystal Palace exhibition, a theme that offers a new perspective on social relations and industrial production in mid-nineteenth-century France. It suggests the importance of consumption in determining social hierarchies and class distinctions and in providing a framework for gender relations among bourgeois men and women. It also helps explain the persistence of various hand methods of manufacturing in France, both in terms of satisfying consumer demand and in terms of buttressing a bourgeois-dominated political structure. This analysis presumes that taste is less universal and aesthetic than historical and social; thus it is useful as a means for understanding social structures and social relations in other countries and at other times. But let us begin with the bourgeois notion of good taste in France at the time of the first international exhibition of industry.

Part One

Consumption in
Mid-Nineteenth-Century France

Chapter One

Constructing the Bourgeoisie through Consumption

The most common response of French women and men to the Crystal Palace exhibition of 1851 was profuse praise for the unrivaled good taste of French manufactured goods, particularly in comparison with those from England. The political economist Adolphe Blanqui asserted: "The main result of the exhibition for the French is the universal, absolute, uncontested recognition of their superiority in matters of art and taste."[1] Only two weeks after the opening of the exhibition, a reporter for the weekly *La Semaine* stated that in the plastic arts France had won the palm of victory at the Crystal Palace.[2] Another French journalist contended that "French superiority in artistic industries is no longer in doubt, even in the eyes of the English."[3] Indeed, British commentators, though sometimes skeptical about the quality of French taste as manifested in manufactured products, acknowledged its widespread influence upon the design and styles of goods from various parts of Europe and North America.[4] The frequent references to good taste in French accounts of the exhibition amount to almost an obsession, yet scholars have largely

1. Adolphe Blanqui, *Lettres sur l'Exposition universelle de Londres* (Paris: Capelle, 1851), 107.
2. *La Semaine,* 16 May 1851, 308.
3. *Le Musée des familles,* July 1851, 319.
4. Ralph Nicholson Wornum, "The Exhibition as a Lesson in Taste," in The Art Journal, *The Art Journal Illustrated Catalogue: The Industry of All Nations, 1851.* London: J. Virtue, 1851. Reprinted as *The Crystal Palace Exhibition Illustrated Catalogue* (New York: Dover, 1970), vi***. See also Tobin Andrews Sparling, *The Great Exhibition: A Question of Taste* (New Haven: Yale Center for British Art, 1982).

ignored this concern in their analyses both of the exhibition and of French industry at that time.[5]

What did "good taste" mean to mid-nineteenth-century French bourgeois, and why was it so important? What insights into the bourgeoisie and into French social conditions in general can an analysis of bourgeois taste reveal? How did the exhibition affect this particular version of good taste? Answers to these questions come from an examination of exhibition reports by the French jury, newspaper accounts of the exhibition, articles in the feminine press, and novels. Almost all of the writers of these works were bourgeois—educated professionals who valued the talent, ambition, and economic accomplishments of their class. Their audience was also bourgeois—a literate public interested in industrial development and in French politics, culture, and entertainment. Most commonly discussions of taste appeared in the feminine press because French bourgeois generally agreed that taste and consumption lay within the purview of women (see chapter 2). But on the occasion of the exhibition men, too, readily ventured to judge the tastefulness of manufactured goods on display in the Crystal Palace. From these materials it is possible to reconstruct the constituent elements of the bourgeois concept of good taste. But how should one interpret these pronouncements, and what can such an interpretation contribute to the understanding of French participation in the Crystal Palace exhibition?

Historical and structural conceptualizations of consumer tastes suggest that specific social, political, and economic conditions underlie any particular set of taste criteria. Writing during the Gilded Age in the United States, the economist Thorstein Veblen noted that social and economic changes contributed to a new pattern of consumption and taste among the newly rich. These people indulged in unrestrained purchases of luxury goods and the ostentatious display of female idleness, according to Veblen, as a means of approximating leisure-class status. Since the newly rich acquired wealth through commercial or industrial activity, they could not completely adopt the refined manners and cultivated tastes of a true leisure class that inherited wealth. They therefore proclaimed their class status through

---

5. A notable exception is Émile Levasseur, *Histoire des classes ouvrières et de l'industrie en France de 1789 à 1870*, 2d ed. (Paris: A. Rousseau, 1904) 2:522–25, 535–77.

what Veblen termed conspicuous consumption—buying expensive and superfluous items that blatantly advertised great wealth.[6]

In contrast to this view—that a new standard of consumer taste and behavior accompanied the economic and social ascendancy of the bourgeoisie in the late-nineteenth-century United States—the sociologist Pierre Bourdieu has more recently explained variations in taste as part of social structuring. In his analysis of France in the 1970s, Bourdieu maintained that factors of occupation, income, and education contributed to social divisions that manifested themselves in different tastes in food, art, leisure activity, home decoration, and so on. In addition, these tastes reproduced themselves in the younger generation as the family, home environment, and neighborhood habituated children to the tastes of the group into which they were born, and rendered alternative tastes strange and undesirable. Though Bourdieu suggested that more education and income could change a person's taste, his concept of taste as a component of social structuring is static and does not readily accommodate change on either the individual or the social level.[7]

What Veblen's and Bourdieu's works indicate is that taste and consumption are not purely personal, aesthetic, or universal; they are also social, political, and mutable. They merit, even require, examination of their social, economic, and political origins in order to understand better the social group adhering to a particular standard of taste and the historical context of that standard. Thus, inquiring into the bourgeois notion of good taste in mid-nineteenth-century France should lead to an investigation of class relations and of the economic and political circumstances that informed them. The following analysis, based on French accounts of the Crystal Palace exhibition, argues that the bourgeois standard of good taste reflected this class's quest for ruling-class legitimacy and for distinction from both the aristocracy and the workers. It also argues that the Crystal Palace

6. Thorstein Veblen, *The Theory of the Leisure Class* (New York: Macmillan Co., 1899).

7. Pierre Bourdieu, *La Distinction: Critique sociale du jugement* (Paris: Éditions de minuit, 1979). For a critique of both Veblen and Bourdieu, see Daniel Miller, *Material Culture and Mass Consumption* (Oxford and New York: Basil Blackwell, 1987). For further analyses of taste and consumption see Arjun Appadurai, ed., *The Social Life of Things: Commodities in Cultural Perspective* (New York: Cambridge University Press, 1986) and Chandra Mukerji, *From Graven Images: Patterns of Modern Materialism* (New York: Columbia University Press, 1983).

exhibition furthered this process by demonstrating the existence of other taste standards and the challenge they represented to bourgeois social, political, and economic domination in France in the context of free-market capitalism and liberal government.

THE USES OF GOOD TASTE

The Crystal Palace exhibition made clear that manufactured goods were increasing in number and variety, and so were consumers. This expansion of consumption represented great opportunities and certain difficulties for the bourgeoisie in France. In the middle of the nineteenth century bourgeois consumers could more readily acquire the styles and furnishings that in earlier times were accessible almost exclusively to the aristocracy. However, few bourgeois had incomes sufficient to buy the finest products of master craftsmen, such as were displayed prominently in the Crystal Palace. There was, then, something of a discrepancy between the exhibition's appeal to popular consumption and the extraordinarily costly items that were the pride of French manufacturing. Writers for the bourgeois press overcame this difficulty with a vocabulary of descriptive terms that both acknowledged the particular condition of bourgeois consumers and allowed them to share with wealthy elites an appreciation for expensive works of art.

Consider Émile Berès's account, in the widely read periodical *L'Illustration,* of a silver dressing table by the silversmith Froment-Meurice (see fig. 5). This magnificent piece of furniture was commissioned by a group of Legitimist women for the exiled Bourbon princess, the duchess of Parma. The dressing table was almost priceless; it was made entirely of silver and crafted and decorated by several prominent artists and artisans. Berès indicated that the true value of the piece lay less in its "richness"—meaning raw material—than in its "art"—referring to the producers' skill. He went on to suggest that any "elegant young woman" could enjoy such furnishings in her dressing room by seeking "art rather than richness" in manufactured goods.[8] A writer for *La Semaine* used essentially the same expression to refer to the superiority of French jewelry over that produced in England. Commenting on displays at the exhibition, he

8. *L'Illustration,* 12 July 1851, 27.

criticized the English for preferring size and weight over workmanship, in contrast to the French tendency to emphasize art and ornamentation rather than the quantity of precious metal.[9] Some years later, Emma Faucon, contributing to the feminine press, repeated the notion in advising readers to furnish their homes with "simple" works of art rather than goods made of expensive materials. "It is not luxury that presided over the decoration of my abode; on the contrary, all is simple. . . . What does the substance [of furnishings] matter to me, provided that my eyes rest with pleasure upon some work of taste, on some product that reflects the talent of a worker or the merit of an artist?"[10]

Terms like "simple," "elegant," "delicate," "charming," "harmony," "purity," "perfection of detail or finishing," as well as "tasteful," were ubiquitous in descriptions of the most costly and heavily decorated French products on show in the Crystal Palace. They invited the bourgeoisie to consider exhibition masterpieces as models for their own, less extravagant purchases of home furnishings and articles of dress. Such expressions suggested a set of consumer values distinct from those of the aristocracy: more restrained, less opulent— but nonetheless discriminating and elitist. Indeed, Constance Aubert, in an essay on home decorating that appeared in *L'Illustration* during the same year as the exhibition, made a virtue of the bourgeoisie's more limited income in contrasting the consumer objectives of the bourgeoisie and of the aristocracy.

According to Aubert, the almost unlimited wealth of the aristocracy and their indulgence in luxury often led aristocratic men and women to exercise poor judgment in consumption. Too much money, she contended, was responsible for aristocrats' seeking personal distinction or originality through buying goods that were shockingly innovative or bigger, more elaborate, and more expensive than what their rivals possessed. Bourgeois consumers, by contrast, were generally restrained from such practices by limited funds. "Especially when immense resources are lacking, intelligent taste must seize the means at its disposal, and must know how to arrange all that decorates and embellishes the residence."[11] "Tastefulness," then, was

9. *La Semaine,* 31 May 1851, 340.

10. Emma Faucon, *Voyage d'une jeune fille autour de sa chambre: Nouvelle morale et instructive* (Paris: Maillet, 1860), 22–23.

11. *L'Illustration,* 18 January 1851, 46.

a substitute for "profligacy" that positively distinguished bourgeois from aristocratic consumers. It also separated them from consumers with even lesser incomes.

Aubert and other writers were equally keen in counseling readers not to buy deceptively impressive articles at a low price, a tendency they associated with less cultivated and poorer consumers. Warning readers against purchasing cheap and gaudy goods, one writer asserted: "In general, everything that is flashy must be avoided. . . . Sensitive women, who are truly economical and have good taste, will flee from [the] display of false luxury."[12] Similarly, on the topic of textiles, another writer maintained that "the appearance of poor-quality fabrics (*des étoffes mauvaises*) is an abuse of industry whose results are often striking." These results, of course, were false economies, when women bought deceptive cloth to save money but looked vulgar and tawdry dressed in such "economical" fabrics. "The woman who is well groomed . . . does not try at all to surprise by the false and showy."[13]

All such expressions as "simple," "elegant," "pleasing," "well designed," and "tasteful" were really code words for the boundaries of bourgeois consumption. By describing products as tasteful, commentators appropriated them for bourgeois appreciation and aspiration, despite the fact that most of the goods displayed in the Crystal Palace by French manufacturers cost far more than ordinary consumers could ever afford. Bourgeois men and women would never buy a Fourdinois buffet, a Rudolphi bracelet, or a Froment-Meurice dressing table, but they could purchase a sideboard that was "simple," jewelry with more "art" than "mass," and furnishings that were modest and imitative of recognizable styles rather than original creations of artists for rich patrons or the state.

Additionally, French commentators used these terms to connote political and national differences. When they were describing manufactured goods, "tasteful" meant "French," in the same way that "cheap," "heavy," "pretentious," or "monotonous" referred to English products. French writers on the exhibition conveyed their notions of national civilizations through the adjectives they used in

12. *Le Conseiller des dames,* May 1850, 218–19.
13. *L'Illustration,* 6 January 1844, 302.

connection with different countries' displays. A certain tension is obvious in the insistence upon France's capacity for up-to-date technology—its leading role as a modern, industrializing nation—alongside the more frequent references to the artistic tradition of French manufacturing. France was indeed industrializing differently from England, and the Crystal Palace exhibition was an opportunity for French industrial experts and politicians to make this difference a positive, national distinction. The repeated use of the term "tasteful" to describe French manufactured goods bridged the gap between aristocratic tradition and bourgeois progressivism. It also connoted the specific characteristics of consumer goods that appealed to the concerns and values of the bourgeoisie in the middle of the nineteenth century: historical styles, extensive ornamentation, heavy padding and drapery, and domestic comfort.

STYLE AND ORNAMENTATION

French men and women who wrote about taste at the time of the exhibition were fairly consistent in their admiration for certain historical styles of furnishings. Renaissance style, for example, was extremely popular among French consumers and was very much in evidence among displays of manufactured goods from France at the exhibition. Equally attractive, according to consumers, jury members, journalists, and writers, were Gothic, Louis XIV, and Louis XV styles. Indeed the height of fashion at this time was to decorate each room of an apartment in a different historical style.[14] The wealthiest of consumers might even procure real antiques for this purpose, but such pieces were rare and terribly expensive. Instead, most bourgeois families bought furniture made by nineteenth-century cabinetmakers imitating the old styles.[15] Critics have deplored this practice as manifesting a lack of imagination and a decline in craft traditions among cabinetmakers, and a deficiency of true taste and originality among

14. *La Gazette des salons,* 7 November 1838, 984; *Le Conseiller des dames et demoiselles,* September 1851, 348–49; Theodore Zeldin, *Taste and Corruption,* vol. 4 of *France, 1848–1945* (Oxford: Oxford University Press, 1980), 74.

15. Zeldin, *Taste,* 74–77; Adeline Daumard, *La Bourgeoisie parisienne de 1815 à 1848* (Paris: SEVPEN, 1963), 136, asserts that most married couples bought new furniture rather than inheriting old pieces; *Le Musée des familles,* October 1851, 31.

consumers.[16] But merely to condemn a society for failing to create a distinctive style is to ignore the reasons behind its imitation of older styles.

Why should styles so disparate as spiritual Gothic, stately Renaissance, ornate Louis XIV, and delicate Louis XV all appeal to bourgeois consumers? Probably because they were all old and were associated with periods of French history when social hierarchies seemed stable and the authority of ruling elites seemed unquestioned. Styles from the glory years of feudal lords, Renaissance princes, and absolute monarchs may have attracted a nineteenth-century bourgeoisie still insecure in their fairly new and hard-won ruling-class status. Though the French Revolution removed the shackles that had inhibited bourgeois men from attaining positions of power, aristocrats and workers continued to challenge bourgeois leadership throughout the first half of the nineteenth century. Moreover, in 1851 this aspiring ruling class had not yet fully developed its own culture or style.[17] It is likely that in adopting the styles of ruling elites during periods when their power was expanding, bourgeois consumers were trying to legitimize their own replacement of (or amalgamation with) the aristocracy as the ruling class of France. For them, the tastes and trappings of power signified power itself. This was evident also in the bourgeois predilection for ornamentation in furnishings, a tendency that was consistent with the preference for the old styles mentioned

16. Lee Shai Weissbach, "Artisanal Responses to Artistic Decline: The Cabinetmakers of Paris in the Era of Industrialization," *Journal of Social History* 16 (Winter 1982): 67–81; Zeldin, *Taste,* 77–79; Mario Praz, *L'Ameublement: Psychologie et évolution de la décoration intérieure* (Paris: Tisné, 1964), 350–51; Nikolaus Pevsner, *High Victorian Design: A Study of the Exhibits of 1851* (London: Architectural Press, 1951). It is worth noting here that in his contribution to the *Encyclopédie,* "The Art of the Joiner," André Jacob Roubo (1739–91) already—in a period considered the zenith of French cabinetmaking—"reproached eighteenth-century craftsmen with being slaves to routine, and for their lack of imagination." *French Cabinetmakers of the Eighteenth Century* (New York: French and European Publications, 1965), 326.

17. Arno Mayer, *The Persistence of the Old Regime: Europe to the Great War* (New York: Pantheon, 1981), esp. chaps. 2 and 4; Rosalind H. Williams, *Dream Worlds: Mass Consumption in Late Nineteenth-Century France* (Berkeley: University of California Press, 1982), 49–50, 108–10; Charles Morazé, *The Triumph of the Middle Classes,* trans. George Wiedenfeld (Cleveland: World Publishing Co., 1966), 125; Roger Magraw, *France, 1815–1914: The Bourgeois Century* (New York: Oxford University Press, 1986), 51–68.

above and that connoted leisure, discrimination, and hierarchical social relations.

In general, the amount and kind of ornamentation on furniture and furnishings distinguished articles as modest or luxurious. A writer for the feminine press explained that a consumer could transform simple furnishings into a sumptuous interior through the addition of ornamentation—by "changing the upholstery fabric, having gilded bronzes of more artistic workmanship, and adding knickknacks (*curiosités*)."[18] The level of simplicity or luxury in a household should, according to this writer, correspond to the wealth and social status of the family: "Let us have the courage of our positions. We shall be simple if that is all we can afford. We shall be rich, surrounding ourselves with luxury, if that weighty task is imposed upon us."[19] Choosing luxurious furnishings was indeed a weighty task, since consumers—usually women—had to distinguish between the vulgar and the tasteful among highly ornamented objects. For example, an expensive mantel clock usually included sculpted figures of gilded bronze; but gracefully arranged figures from classical mythology might be more tasteful than a heavy allegorical representation, depending on the style of the room for which the clock was intended and on the quality and originality of the workmanship.

As with old styles, consumers probably favored highly decorated items because ornamentation represented both wealth and social distinction. Ornamentation was a clear sign of wealth because it used expensive materials like gold, silver, precious gems, and exotic woods. Ornamentation also reflected the skill, creativity, and labor of craftspersons—and the consumer's cultivation and discrimination in selecting works of art, not just expensive furnishings. Bourgeois consumers considering the purchase of, say, a sculpted Gothic chair or a chased Louis XV tea service could feel like patrons of the arts judging the ability, taste, and amount of time that an artisan put into an original piece. In this way, they imaginatively reproduced the patron/artist relationship of the old aristocracy and the craftspeople they commissioned. To be an art patron, as opposed to a mere consumer of goods, required esoteric knowledge and appreciation

18. *Les Modes parisiennes*, 22 March 1851, 3009.
19. Ibid.

that only persons of wealth and cultivation could acquire.[20] As with the popularity of old styles, then, the taste for ornamentation linked bourgeois consumers with their aristocratic predecessors as members of the consuming, discriminating, and therefore ruling class.

In the nineteenth century the ornamentation of furnishings often required artisan skill in the application of new techniques as well as in the carrying on of older methods of hand manufacturing. While bourgeois consumers sought the personal touch of a worker's hand and mind, they also appreciated producers' use of hand machines and scientific methods that constituted a form of industrial progress. Small hand tools often permitted workers to reproduce patterns with greater precision than with the unaided hand. Louis Wolowski, a jury member who reported on furniture at the exhibition, asserted that "great perfection, in the smallest details of execution," was an essential part of good taste.[21] The dynamism that many art historians discern in mid-nineteenth-century decorative arts derived from a certain pride in modern technology and from producers' willingness to exploit that technology in the production of "old style" articles.[22] There was also an exuberance in the bourgeoisie's enjoyment of completely decorated surfaces; the innovative decoration of old styles combined the new and the traditional in consumer goods.

The effect of consumer demand upon production processes is discussed in more detail in part 2; selected examples from French exhibits at the Crystal Palace will serve here to illustrate the multiple meanings of ornamentation to mid-nineteenth-century consumers in France. An earthenware platter made by the potter Charles Avisseau won praise and a prize medal at the exhibition for its originality, tastefulness, and technique. The platter, totally covered with brightly colored marsh fauna and flora—lizards, snakes, fish, snails, leaves, seaweed, and lily pads—was useless as a serving dish. The numerous

20. Bourdieu, *Distinction;* Brian Spooner, "Weavers and Dealers: The Authenticity of an Oriental Carpet," in *Social Life of Things,* ed. Appadurai, 195–235.

21. Commission française sur l'Industrie des Nations, *Exposition universelle de 1851: Travaux de la Commission française sur l'Industrie des Nations* (Paris: Imprimerie impériale, 1855) 7:2.

22. Philadelphia Museum of Art, *The Second Empire, 1852–1870: Art in France under Napoleon III* (Philadelphia: Philadelphia Museum of Art, 1978), 15; John M. Hunisak, "Beyond Pomp and Circumstance: Another Look at Second Empire Art," *Art in America* (January–February 1979): 79–83.

sculpted and glazed figures on the platter left no room for food or anything else; the object was clearly intended for display only. However, as imitations of nature the platter decorations are remarkable. They attest to Avisseau's careful study of plants and animals and to his ability to design and execute an extremely complex and detailed scene in three dimensions. Moreover, Avisseau had perfected several new techniques in glazing that enhanced the quality of the colors of the decorated piece. To modern eyes this platter may appear over-decorated, loud, and in utter violation of the principle that form should be related to function. However, French commentators found Avisseau's platter supremely tasteful because it was richly orna-mented; it represented enormous skill, time, and originality on the part of the producer; its resemblance to nature was readily apparent and appreciated; and it was technologically a tour de force.[23]

The sideboard by the cabinetmaker Fourdinois (fig. 6) was also highly decorated and skillfully executed, though it differed from Avisseau's platter in adopting a distinctive historical style. This enor-mous work of carved walnut took its inspiration from the Renais-sance, displaying variations on the scrolled shields and tracery char-acteristic of that period. Its function is suggested by the carved figures relating to food and drink, though it is difficult to determine from the illustration just how this piece of furniture could hold food or dishes. Holding up the sideboard are six chained, seated hounds, suggesting the theme of hunting as a means of procuring food for the table. This theme is repeated in higher layers of carvings, where a felled deer and game bird appear, as well as scenes of hunting and fishing. Above these representations stand four female figures sym-bolizing Europe, Asia, Africa, and America and holding fruits or products typical of each continent. At the top of the sideboard six putti are harvesting grain and making wine; between them sits the goddess Ceres holding two cornucopias. Journalists and jury mem-bers were impressed with this intricate design and appreciated its appropriateness to the function of the sideboard. This appropriateness made the Fourdinois masterpiece tasteful, in contrast to the many exhibits whose designs were incongruous with their intended func-tions, such as a silver cream ladle in the shape of a buttercup, or a silver vulture on the lid of a perfume bottle. Critics also praised the

---

23. Avisseau is discussed further in the last section of this chapter.

sideboard's design for being original and imaginative while still true to the historical Renaissance style.[24] Though few bourgeois consumers could actually buy such an elaborate and costly sideboard, it was an exemplar of tasteful ornamentation and style.

COMFORT AND INSULATION

An essential component of the bourgeois definition of good taste at the time of the exhibition was the comfort and suitability of furnishings. In his report on furniture at the exhibition, Wolowski wrote that "the piece of furniture must lend itself easily and without obstruction to the use it serves." He roundly condemned beds that were so monumental that no one could get a good night's rest in them, chairs so covered with ornamentation that people bruised their bodies when trying to sit in them, and tables that tore people's clothes when they passed.[25] Madame Pariset, the author of a popular housekeeping manual revised and reprinted several times through the century, echoed Wolowski's call for accommodating furniture, and she added stipulations for harmony and practicality. She advised her female readers to select furnishings that were "useful, convenient, durable, and especially that go together well."[26] In principle, even bourgeois families of modest means could follow these simple rules of tastefulness in home furnishing and could thereby guarantee a modicum of comfort as well.

Comfort was a key word in mid-nineteenth-century pronouncements on taste and furnishings, for it was closely linked with an attachment to the home and an ideal of domestic happiness particular to the bourgeoisie of this period. Aubert defined the comfortable as "all that relates to daily usage. It is not luxury, it is not whimsy, nor is it objects of absolute necessity. It is the thousands of resources of which well-being and good living (*savoir-vivre*) consist."[27] Aubert was explicit in distinguishing the criterion of *comfort*, characteristic of the bourgeoisie in the nineteenth century, from *luxury*, which she maintained motivated aristocratic consumption in earlier times. For

24. Wornum, "Exhibition," xiii***.
25. Commission française, *Exposition* 7:2–3.
26. Mme Pariset, *Nouveau Manuel complet de la maîtresse de maison* (Paris: Roret, 1852), 10.
27. *L'Illustration*, 18 January 1851, 46.

Aubert, consumers who furnished their homes and dressed themselves with the goal of comfort in mind were exercising good taste; comfort and taste were interchangeable, according to Aubert, where bourgeois consumption was concerned. Moreover, she maintained that contemporary bourgeois consumers who adhered to her principles were more tasteful than the preceding consuming class—the nobility—because the former eschewed luxury. "Taste is superior to luxury, because what pleases is superior to what surprises," Aubert asserted.[28]

Similarly, the economist Émile Berès, also writing for *L'Illustration,* echoed Aubert's association of good taste with domestic comfort. In an article on the Crystal Palace exhibition he praised the nineteenth-century involvement in the home as an indication of high civilization: "One cannot be too happy seeing man seek, according to his means and in purifying his tastes, to give charm to his [household] interior. This taste is intimately linked to family happiness, the first, the most certain, as well as the most desirable of all tendencies."[29]

It is noteworthy that Berès referred to "man" in this context of home decoration. He may well have been using the term to represent humankind in general, for (as discussed later in this section) both men and women of the bourgeoisie concurred that women should be arbiters of taste and primary consumers for the home. But it is also likely that Berès quite deliberately encouraged male participation in domestic activities because the home, though commonly seen as a female or private sphere, was an essential foundation for the male-dominated, bourgeois social order of the nineteenth century.[30] Bourgeois men could freely engage in productive, commercial, administrative, and political activities outside of the home precisely because women provided them with a "haven" from the stresses of public life at the end of the workday, and more importantly, women's domestic duties of housekeeping and child care ensured the reproduction of the social hierarchy with future generations. In addition, the home permitted bourgeois men and their families to evade the unpalatable conditions of social inequality and working-class poverty that were

28. Ibid.
29. Ibid., 19 July 1851, 39.
30. See also Leonore Davidoff and Catherine Hall, *Family Fortunes: Men and Women of the English Middle Class, 1780–1850* (Chicago: University of Chicago Press, 1987).

ubiquitous during the early stages of industrial capitalist development.

Consider the following passage by Constance Aubert describing the ideal bourgeois living room.

> [It is] well enclosed by good door curtains, by cushions of silk, and by double draperies that hermetically seal the windows. There may be only a paper wall covering, but a good rug is underfoot; people sit in excellent small seats without roughness, where the body abandons itself and rests. A profusion of fabric adorns the windows, covers the mantel, hides the woodwork. Dry wood, cold marble, disappear under velvet or tapestry.[31]

Aubert went on to describe the writing desk, plant stands, and other furnishings necessary for the "material well-being" that she considered the hallmark of bourgeois comfort. The meaning of "comfort" in this instance was the elimination of all hard surfaces and sharp edges. The perfect housewife (Aubert explicitly assigned furnishing to the mistress of the home) disguised or hid real structures to make them soft, accommodating, and restful. Why should bourgeois householders be so intent upon hiding wood and marble beneath upholstery and fabric and "hermetically sealing" the household interior from the outside? To be sure, the door curtains, thick rugs, and double curtains sealing the windows protected the bourgeois family from the cold; and as chapter 3 will demonstrate, Parisian householders did indeed cover their windows, floors, and furniture with layers of fabric and padding. But it is possible to view the bourgeois penchant for insulation in the home as an effort to insulate the family from a reality much harsher than wood and marble.[32] After all,

31. *L'Illustration*, 15 February 1851, 112.

32. There is remarkable similarity between the following analysis by the twentieth-century critic Walter Benjamin and Aubert's prescriptions for tastefulness in the bourgeois home. Benjamin interpreted the domestic interior of the July Monarchy as a deliberate effort on the part of the bourgeois householder to shut out the reality of social problems and to create a fantasy world of security, isolation, and comfort: "The private person who squares his accounts with reality in his office demands that the domestic interior be maintained in its illusions. This need is all the more pressing since he has no intention of extending his commercial considerations into social ones. In shaping his private environment he represses both. From this spring the phantasmagorias of the interior. For the private individual the private environment represents the universe." Walter

Aubert wrote only three years after the working class had risen against the bourgeois state, the free market, the sanctity of private property, and the ethic of individualism. Though military force successfully repressed the proletarian challenge, Aubert and members of her class had reason to try to distance themselves from the unsolved problems of poverty and the radicalism it inspired. A comfortable apartment or house that created a space for the bourgeois family, separate from the public world of the marketplace and the streets, allowed for a sense of insulation from poverty, hostility, and violence. It also fostered illusions of bourgeois security and self-sufficiency, as in the disguising of hard surfaces and the barricading against the outside.

The thick drapery, cushioned sofas, and curtained windows, as well as ornamented furniture, sharply distinguished bourgeois homes from those of workers. In general, working-class furnishings consisted of only the most basic items: beds, bedding, cookware, a table and chairs, and perhaps a wardrobe for clothes or a buffet for dishes. Since housing was scarce, and rents rose steadily throughout the nineteenth century, the habitations of workers were notoriously dilapidated, overcrowded, and unhealthy. Small wonder that much working-class family life and sociability occurred in public—in the streets and cafés—instead of in domestic privacy.[33] Interior comfort and style were amenities that few workers could afford and that bourgeois families therefore prized all the more.

ALTERNATIVE STANDARDS OF TASTE AT THE EXHIBITION

Defending the bourgeois standard of stylish ornamentation and domestic comfort was a political issue in the broad sense of the term, as

---

Benjamin, "Paris, Capital of the Nineteenth Century," in *Reflections: Essays, Aphorisms, Autobiographical Writings,* trans. Edmund Jephcott (New York: Harcourt Brace Jovanovich, 1978), 154. See also Michelle Perrot et al., *De la Révolution à la Grande Guerre,* vol. 4 of *Histoire de la vie privée,* ed. Philippe Ariès and Georges Duby (Paris: Éditions Seuil, 1987), 308–23.

33. Perrot et al., *De la révolution,* 314–19, 361; Frédéric Le Play, *Les Ouvriers européens,* 2d ed. (Paris: E. Dentu, 1878) 6:327–492. For a controversial interpretation of the wretchedness of workers and their miserable living conditions, which led to criminal behavior, see Louis Chevalier, *Laboring Classes and Dangerous Classes in Paris during the First Half of the Nineteenth Century,* trans. Frank Jellinek (New York: Howard Fertig, 1973).

French commentators' negative responses to alternative demand criteria at the Crystal Palace exhibition made clear. Inside the Crystal
Palace, French journalists and jury members noted many manufactured articles, especially from England but including some from
France, that were deemed in poor taste or of inferior quality. In
several instances, commentators blamed consumers for this problem,
noting that manufacturers who catered to markets other than the
French bourgeoisie risked violating the bourgeois standard in order to
sell their wares. These writers discerned a direct link between consumer tastes and production processes, indicating that consumers
who lacked the good taste typified by the bourgeois standard encouraged the manufacture of showy, cheap, and poor-quality products.

Émile Berès, for one, was not particularly worried about the
alternative demand criteria he noticed in 1851. In an article on French
bronze making at the exhibition he criticized the products by a
manufacturer named Miroy, asserting that "in Monsieur Miroy the
genius of art is less predominant than the mind of the industrialist."
Berès's explanation for the poor quality of Miroy's bronzes was the
manufacturer's profitable trade with North and South American
consumers whose tastes were "less cultivated, less exigent than our
own." Berès went on to suggest how a certain type of demand forced
manufacturers into compromising the bourgeois standard of good
taste:

> The inhabitant of the Pampas, the planter on the banks of the Ohio,
> the Mexican Sybarite, the dashing girl of Havana want ROCOCO,
> POMPADOUR; give them rococo and pompadour. The industrialist is
> not exactly a moralist, a philosopher; nor is he . . . a Benvenuto or a
> David. The CASHBOX has the most reverberating sound for him; let us
> not blame him for listening to it.[34]

In Berès's opinion, rococo and pompadour styles were in bad taste
because they did not contribute to comfort in the home; they represented a period of aristocratic decadence inimical to the domestic and
conformist values of nineteenth-century bourgeois. Berès protested
that he understood and accepted Miroy's justification for betraying
the bourgeois standard of good taste—entrepreneurial profit—but he
also refused to admit that the poor taste of American consumers, and

34. *L'Illustration*, 19 July 1851, 39.

commercial exigency in general, could dethrone the standard he espoused. "France . . . is rich enough in ideas and flexible talents to be able to respond at will to all needs, to all fantasies."[35]

For Berès, then, the bourgeois standard was the ultimate and universal measure of tastefulness, and not yet seriously threatened by alternative demand types. Nor did Berès express concern over Miroy's choice of profit over quality in his manufacturing operation. The implication was that bourgeois consumers in France would continue to uphold their class standard in demanding high-quality manufactured goods, despite the appearance of new markets and the lure of success in accommodating them. Other commentators, however, were less sanguine than Berès about the implications of alternative demand criteria, especially for the French economy and social stability.

Two French commentators responded to the comparison of French and English industry at the Crystal Palace exhibition by berating English standards of consumption and production in favor of the French bourgeois standard, and implicitly of manufacturing methods in France. The baron Charles Dupin, head of the French jury and a respected statistician, contended that both consumers and producers in Great Britain considered profit more important than quality in manufacturing, with a negative effect on the production of British textiles.

> What did it matter to English and Scottish [consumers] whether [manufacturers] wished to spin a fine thread or weave a beautiful cloth! That they make a fortune if they can . . . will be their reward. But as for exhibitions, if they notice one spinner or weaver rather than another, the latter can pride himself on selling more than his competitors, which would sadden them.[36]

Clearly Dupin did not consider "making a fortune" to be the sole nor the best objective of manufacturing, and he was disturbed that this predilection on the part of British consumers and producers undermined the purpose and criteria of exhibitions. For the jury of the 1851 exhibition judged products primarily on the basis of design, utility, originality, and craftsmanship—criteria that corresponded

35. Ibid.
36. Commission française, *Exposition* 1:76.

closely to the elements of bourgeois taste.[37] By these criteria French manufactured goods showed well at the exhibition and gained world-wide recognition. However, in terms of profitability and efficiency, British manufacturing far outstripped its continental rivals.[38] If such criteria were to replace the bourgeois standard, French manufacturing would compare much less favorably with that of Britain in world markets. Like Berès, Dupin did not express any immediate alarm at this prospect, but the exhibition did serve notice on France that not all consumers in the world sought the quality goods that distinguished French manufacturing. In economic terms, Dupin and Berès were correct in discerning no danger to the ability of France to dominate its chosen market in 1851. But in social and political terms, demand for cheap, mechanically made, standardized goods was more threatening to the bourgeoisie and the social order they upheld.

According to French periodicals, British writers suggested that the French superiority in the manufacture of luxury goods belied France's pride in its democratic institutions and society.[39] If France were truly democratic, the argument ran, then it should produce more low-priced goods for the working poor. The implication here was that Britain, with its factories and mass production of low-cost consumer goods was more "democratic" than France, where most workers engaged in the manufacture of luxury products they could never hope to buy. At least one French commentator considered this criticism to

37. Official exhibition criteria for judging certain categories of goods were as follows: "Timepieces—strength, durability, simplicity and economy of construction, the finish of workmanship in accordance with scientific conditions of execution; manufactured goods (fabrics)—increase in usefulness and new usages, superiority of quality and workmanship, design, taste, low cost; glass and porcelains—great usefulness combined with economy and elegance." *La Semaine,* 25 July 1851, 467. French jury members made much of the fact that the English organizers of the exhibition devoted the bulk of the proceeds to the establishment of art and design schools in London. Ministère de l'Intérieur de l'Agriculture et du Commerce, *Annales du commerce extérieur: Faits commerciaux,* no. 20, *Exposition universelle de Londres en 1851* (Paris: Imprimerie impériale, 1853), 25–26; Léon Emanuel de Laborde, *De l'union des arts et de l'industrie* (Paris: Imprimerie impériale, 1856) 1:390–91.

38. See the French jury report on French and English glassmaking, where the reporter extolled the quality of French glass while acknowledging the more efficient methods of production and marketing of English glassmakers. Commission française, *Exposition* 6:22, 30; Michel Chevalier, *L'Exposition universelle de Londres* (Paris: Mathias, 1851), 17, 24–25, 34; *La Semaine,* 31 May 1851, 339.

39. *Le Musée des familles,* July 1851, 319; *Le Correspondant,* 9 June 1851, 307–8; *La Semaine,* 13 June 1851, 372.

be a rude insult to the French bourgeoisie and a sly encouragement of working-class unrest in France. A journalist for *Le Musée des familles* exclaimed:

> We reply to John Bull that bread is still cheaper in Paris than in London. Charity in France gives three million daily for the poor. Our most miserable hovels are palaces compared to the slums of Saint Giles, and our luxury workers would be happy as lords if they did not have . . . the chronic illness of revolutions. Cured of it once and for all, they will have nothing to envy about your calico at four sous per meter.[40]

In defending the bourgeoisie in France, and implicitly the bourgeois standard of consumption, this author betrayed his bitterness and hostility toward the French working class and their recent revolution. In his view, further efforts toward mass production methods that would provide the French working classes with cheap manufactured goods were an unnecessary concession to unseemly and destabilizing rebellion. Like other members of his class, he believed that workers should be grateful to the bourgeoisie for demanding luxury goods and so providing jobs for skilled craftsmen and unskilled laborers alike. He resented the suggestion that the middle class in France failed to address the consumption needs of workers; although he mentioned cheap bread and bourgeois charity, he might also have referred to the expansion of the ready-made clothing industry that catered primarily to working-class customers.

Were Berès's, Dupin's and this journalist's responses to the alternative demand criteria evident at the exhibition mere outpourings of patriotism? Were they rationalizations against the obvious superiority of Britain in the production of low-cost goods for working-class consumption and for export? Certainly chauvinism was not absent from any of the French accounts of the exhibition; this is not surprising given the rise of nationalism throughout Europe during the nineteenth century. Moreover, French commentators hoped their country's performance would show that the manufacturing crisis that both preceded and accompanied the 1848 revolution was over, and that foreign and domestic trade was again flourishing. But patriotism

---

40. *Le Musée des familles*, July 1851, 319. See also *Le Correspondant*, 9 June 1851, 307–8.

cannot be regarded as the sole explanation for the commentators' obsession with taste and their denigration of English manufacturing in this regard, for two reasons.

First, the French were not the only viewers to judge French taste superior to that of other developing countries. The entire international jury essentially agreed on this point, and one English writer's report on precious metalwork displayed by England and France agreed with the French commentators as to the positive influence of consumer taste upon manufacturing in France:

> The English gold and silversmiths seemed to have valued their work by its weight of metal, while their foreign rivals wisely considering that art would give a more real and permanent value than more material, seemed to have attempted to attain perfection in design and workmanship. . . . From this we may infer that the blame is not to be attached so much to the producers of English Works of Art, as to those who, as large purchasers, keep the market stocked with articles suited to their own bad taste. . . . It cannot be denied that the taste of the class who purchase these works abroad must be higher than that of the corresponding class in this country.[41]

Thus the French bourgeois were by no means alone in praising the good taste of French consumers and manufacturers.

Second, the patriotism argument cannot be divorced from a political interpretation of the French response to alternative demand criteria. French jury members and journalists, all bourgeois, were defending not only the bourgeois standard of taste in consumption and manufacturing in France but also bourgeois domination of the French economy, politics, and society. In rejecting alternative demand criteria, they reinforced the ascending position of their class. Their attitude was not unlike the attitudes behind the sumptuary laws of earlier centuries that prohibited all but aristocratic elites from wearing certain colors and apparel.[42] To admit that foreigners, or French workers, had legitimate demands as consumers that differed from the bourgeois standard was to concede that certain social needs were not being met. Indeed the suggestion that workers were con-

41. *Dickinson's Comprehensive Pictures of the Great Exhibition of 1851* (London: Dickinson Bros., 1854) (no page numbers).

42. Appadurai, "Introduction: Commodities and the Politics of Value," in *Social Life of Things,* 32, 57.

sumers at all, and that they needed cheaper goods in order to enjoy a decent standard of living, called into question the existing social order in France and the liberal, individualistic ethos that supported it. Addressing workers' consumer needs would imply major changes in the organization of production and distribution of goods in France. This is precisely what workers demanded in 1848, and the bourgeoisie forcefully refused them. Were bourgeois men, either as political leaders or as individual entrepreneurs, willing to acknowledge the working-class demand for cheap consumer goods and the changes in manufacturing this demand entailed in 1851?

Perhaps a minority were. A reporter for the *Moniteur industriel,* a periodical representing the interests of large-scale industrialists, asserted that although good taste in manufacturing was all very well, French producers should be doing more to open new markets for lower-priced manufactured products:

> We are always infatuated with our taste, a very great and precious thing in manufacturing, which has placed us in the high rank we occupy; but I think that we must now concentrate on methods to produce cheaply, to maintain foreign competition, and to put our production further at the disposal of the mass market that we have barely addressed. I believe that we have progress to make in this direction.[43]

Among both official and popular accounts of the exhibition, such judgments on French taste and manufacturing were infrequent. Moreover, this writer's comments could easily be interpreted as self-serving (or at least supportive of the big industry and/or protectionist lobby) more than concerned with the welfare of low-income consumers. For economic reasons most French commentators thought that French manufacturing should do what the exhibition showed it did best—produce tasteful, high-quality consumer goods for domestic and foreign markets (see chapters 6 and 7). Political concerns also squelched widespread advocacy of the production of

43. *Moniteur industriel,* 10 July 1851. Compare this modest and isolated plea for change in French industrial production with the flood of literature in the United States over the past few years urging American manufacturers to adopt Japanese methods of production organization; or even compare it to the English response to the Crystal Palace exhibition and the admission that England needed to improve its art and design in manufacturing.

cheap goods for mass markets because "mass markets" usually meant workers, and workers' tastes were significantly different from those of the bourgeoisie. Indeed, Natalis Rondot's report of clay pipes at the exhibition (discussed in the next section) indicated that workers' "tastes" were downright subversive of bourgeois rule.

TASTE AND SOCIAL RELATIONS

Analyzing clay pipes as part of his report on *articles de Paris,* Rondot bemoaned the inferiority of these products compared to other items displayed by French producers of *articles de Paris* (brushes, combs, decorative boxes, pins, toys, canes, umbrellas, buckles, buttons, etc.). His explanation for this problem was the bad taste of the growing number of working-class consumers, who forced manufacturers to produce cheap and ugly pipes.

> We are the first to regret that this immense consumption of clay pipes does not serve . . . to purify the taste and elevate the ideas of the masses; clearly the busts of great men or lovely statues would be better than these heads of revolutionary heros and broad caricatures. But the manufacturer must account for the habits of consumers, and one cannot hide the fact that the majority of peasants and workers prefer ephemeral figures or ugly models to the most charming subjects that Greek, Arab, or Egyptian art would inspire.[44]

Rondot's report raised two important issues regarding taste and manufacturing in France: taste as a means of political control or subversion, and the conflict for manufacturers between good taste and consumer demand. Rondot had obviously hoped that through consumption the working class would adopt the tastes, and thereby accept the social and political domination, of the bourgeoisie. But if great men and great art on clay pipes could impose bourgeois order, then pipes decorated with revolutionary heros and caricatures could conversely promote social unrest. Here, then, was an important reason for French writers to uphold the bourgeois standard of taste and to laud manufacturers who did so in their products. Though jury members and journalists never actually condemned manufacturers for violating the bourgeois standard and catering to the untutored tastes

44. Commission française, *Exposition* 7:129.

of foreign or working-class consumers, their accounts nonetheless damned with faint praise the French producers who indeed cherished profit over good taste. Did these writers recognize a contradiction between maintaining the bourgeois standard of consumption and producing for a nonbourgeois market? Writers in Britain, where production was often more efficient and more modern than in France, were keenly aware of this problem on the occasion of the exhibition, and they responded to it by actively cultivating expertise in design and art among producers.[45] The issue was not yet critical in France, but the very effort to disguise it suggested some awareness that elite tastes and a free market were not always compatible.

French bourgeois took refuge from the difficulties of acknowledging the working-class consumers' preferences, and especially their needs, by fabricating a historical model of consumer and producer relations based on working-class, artisan production for bourgeois, elite consumption. This archetypal social relationship was exemplified in the biography of the pottery maker Charles Avisseau, who successfully exhibited at the Crystal Palace and who came to represent, for bourgeois readers of the popular press, the ideal French worker-producer. The role of the bourgeois in this feature story was, naturally, that of tasteful consumer. The roles of producer and consumer in this account of the middle of the nineteenth century paralleled the roles of skilled artisan and patron of the arts, respectively, from sixteenth-century France.

Charles Avisseau was born in Tours in 1796, the son of a poor stonecutter who occasionally worked in a pottery establishment when he had no work in his own trade. To get the young Charles out of his mother's way, his father often took him to the pottery works, where Charles imitated the glaze makers. The owner of the works recognized young Avisseau's drawing talent and hired him to work in the shop. Avisseau quickly learned all about clays, firing, and glazing, and he moved on to a supervisory position at a fine faience works in Beaumont-les-Autels. There he initiated improvements in oven construction and clay and mineral mixtures, and he put his own ideas into pottery creations. At this point in his career, Avisseau saw an example of Bernard Palissy's glazed pottery, and he determined to discover the sixteenth-century potter's secret of applying colored

45. Sparling, *Great Exhibition.*

enamel. With no formal education or help, Avisseau succeeded in his goal through dogged experimentation.

Avisseau was driven to further experimentation in pottery coloring. Quitting his job and returning to Tours, he bought his own small shop and made a living by producing church ornaments and statues of saints, as well as repairing plaster. But he spent his nights searching relentlessly for a new palette of pottery colors that would fire at the same temperature. On his own Avisseau consulted learned treatises and studied nature to accomplish his goal, barely keeping poverty at bay during his artistic and scientific researches. Though he was again successful, he was still unsatisfied. He ultimately sought to incorporate gold into his glazes. In melodramatic terms, the biography described the Avisseau family engaged in the quest:

> Around the table . . . father and son, palette knife in hand, continue to work, after a full day, with the naive ardor of sixteenth-century artists. Under their direction two little sisters trace, with the patience of Benedictine monks, the scales of serpents and the veins of leaves modeled by the artists. Near the fire the mother of the family . . . pulverizes the glazes on a small grindstone.[46]

In the climax of this tale of artistic genius at work, the potter had exhausted all of his resources and had no more gold for his final glazing experiment, when his wife offered him her gold wedding band. The driven artist prevailed over the tender husband and honorable man; Avisseau accepted her sacrifice, and eventually produced the precious glaze.

Despite these many personal triumphs Avisseau was unknown as an artisan until 1845, when a lawyer and dilettante bought and displayed one of his decorated bowls. At the lawyer's urging, Avisseau exhibited in local and regional shows, and finally at the 1849 Paris exhibition. The director of the national pottery manufacture at Sèvres, Brogniart, invited him and his family to live and work there on condition that Avisseau divulge the secret of his glazing processes. Acknowledging the honor, Avisseau declined, stating that he preferred the freedom of being his own master. Courted by artists, notables, and royalty he continued to live simply and devoted himself solely to the production of ceramic art. His noteworthy accomplish-

46. *Le Musée des familles,* March 1851, 181.

ments included vivid imitations of animals in their natural habitat, like his award-winning platter at the Crystal Palace exhibition representing swamp creatures and foliage.[47]

This story of Avisseau represented a perfect bourgeois ordering of consumption and production, modeled after the social and economic relations of the Renaissance. Avisseau, a skilled and dedicated artisan, succeeded because of individual effort, talent, and bourgeois patronage. He knew his place in the social hierarchy and aspired only to continue to supply bourgeois consumers with works of art for decorating their homes. The frequent references to the sixteenth-century—Palissy, naive artisans, Benedictine monks—suggested a simpler era, before industrialization threatened skilled labor, created urban slums, and impelled workers toward revolution. Bourgeois readers must have reveled in the presentation of an artisan totally devoted to his craft, who not only was satisfied with his station in life but also flattered the bourgeois desire to fill the sixteenth-century role of aristocratic patronage of the arts. In the story the representative bourgeois figure was not entrepreneurial but professional, and his main contribution to industrial achievement in France was cultivated taste and discriminating consumption. The lawyer and dilettante experienced no conflict between his taste as a consumer and his "productive" role in society. Consumers and producers were clearly divided here, just as they were during the Renaissance; and just as this ordering had contributed to France's reputation of artistic and tasteful manufacturing in the sixteenth century, readers could expect it would continue to do so in the 1800s. The proof was in the universal recognition of the fine French performance at the Crystal Palace exhibition.

CONCLUSION

What the exhibition showed, among other things, was that consumption was both a source and a characteristic of power. Bourgeois consumers promoted a standard of taste that effectively limited the acquisition of durable, stylish, comfortable furnishings and clothing to their own class. Possession of these goods was a visible sign of membership in the elite class, and accounts of the exhibition

47. Ibid., 179–84.

suggested that the bourgeoisie was not eager to dispense with its almost exclusive hold on taste and consumption. French commentators greeted with derision the many displays in the Crystal Palace that reflected the tastes of consumers other than French bourgeois. And they decidedly scorned the suggestion that French manufacturers should address the requirements of a mass market, finding this a threat to bourgeois rule in France and an encouragement of working-class discontent with the existing economic and political order. Though jury members and journalists acknowledged the benefits that accrued to manufacturers who supplied foreign or working-class markets with cheap and "tasteless" goods, they did not view this as a desirable or even an inevitable direction for French industry in the future. Their positive assessment of the superior quality and taste of French manufactured goods reflected both economic and political motivations to maintain the status quo.

Bourgeois taste encompassed aristocratic models of art and beauty, and a new criterion of domestic comfort. The ideal of tastefulness as conveyed in the popular press was a respectable conformity, promoting harmony and style without being extraordinary or garish. Women more than men had to fulfill the goal of tastefulness for themselves and their families, particularly since the element of domestic comfort placed consumption within the feminine domain of the home. Yet the consuming activity of bourgeois women was neither limited to the home nor a simple complement to the male realm of production, despite the idealized gendering of private and public in nineteenth-century middle-class society.

Chapter Two

"To Triumph before Feminine Taste"

Female Consumption, Gender, and Women at the Exhibition

According to Armand Audiganne, a prolific writer and expert on industrial matters, the Crystal Palace exhibition of 1851 was decisive in demonstrating the positive influence of female consumers upon manufacturing in France. He asserted that the comparison of French and British consumer goods in the Crystal Palace revealed the artistic shortcomings of British manufacturing, and he blamed these shortcomings on the failure to consult women's taste or meet female demand. "The habits and usages of English society were fatal for art," Audiganne explained. "They did not permit feminine instinct to shine forth." Writing on the occasion of a later exhibition in 1867 Audiganne reiterated his belief in the importance for industrialists of considering female consumers' judgment: "To triumph before feminine taste is worth more to manufacturing than to succeed before the most thoughtful and least arbitrary decisions of [exhibition] juries."[1]

Audiganne here acknowledged that in the nineteenth century women of the comfortable, if not wealthy, classes assumed a significant role as consumers of manufactured goods. Until very recently, however, scholars have neither questioned this observation nor analyzed its origins and consequences.[2] How did taste and consumption

---

1. Armand Audiganne, *La Lutte industrielle des peuples* (Paris: Capelle, 1868), 187, 186.

2. A notable exception is Thorstein Veblen, *The Theory of the Leisure Class* (New York: Macmillan Co., 1899). Recent works that address the topic of feminine taste and consumption in French history include Debora L. Silverman, *Art Nouveau in Fin-de-Siècle France: Politics, Psychology, and Style* (Berkeley: University of California Press, 1989); Leora Auslander, "Women Subjects and Feminine Objects: Women as Consumers and as Images in Late Nineteenth-

become "feminine" in nineteenth-century France, and what does this reveal about social organization and gender relations during industrialization? In what ways did female consumption support bourgeois and patriarchal social ordering, and also undermine it? And what were the implications of female consumption for industrialization, as presented in the Crystal Palace exhibition? Such questioning introduces a new approach to understanding gender relations and their place in history.

Many influential studies of nineteenth-century bourgeois women have framed the subject within the concept of separate spheres. That is, they have analyzed women in the context of the private realm of the home that contrasted with, and to some extent complemented, the male sphere of public activities like politics, trade, and production.[3] However, the notion of gender, referring to the social and cultural constructs of femininity and masculinity, allows for a more complex and nuanced understanding of male and female roles and relations. Analyzing consumption from the perspective of gender crosses the boundaries of separate spheres for women and men to reveal the public as well as private function of women's role as consumers for their families. It also elucidates contradictions in the nineteenth-century ideal of feminine domesticity by showing how

---

Century Paris" (Paper presented at the Seventh Berkshire Conference on the History of Women, Wellesley, Mass., 19–21 June 1987); Whitney Walton, " 'To Triumph before Feminine Taste': Bourgeois Women's Consumption and Hand Methods of Production in Mid-Nineteenth-Century Paris," *Business History Review* 60 (Winter 1986): 541–63; Michael B. Miller, *The Bon Marché: Bourgeois Culture and the Department Store, 1869–1920* (Princeton: Princeton University Press, 1981). See also, for the United States, Susan Porter Benson, *Counter Cultures: Saleswomen, Managers, and Customers in American Department Stores, 1890–1940* (Urbana: University of Illinois Press, 1986); William R. Leach, "Transformations in a Culture of Consumption: Women and Department Stores, 1890–1925," *Journal of American History* 71, no. 2 (September 1984): 319–42.

3. The literature on feminine domesticity in Western Europe is too vast for complete citation here. Works particularly relevant to this study are Bonnie G. Smith, *Ladies of the Leisure Class: The Bourgeoises of Northern France in the Nineteenth Century* (Princeton: Princeton University Press, 1981); Erna Olafson Hellerstein, Leslie Parker Hume, and Karen M. Offen, eds., *Victorian Women: A Documentary Account of Women's Lives in Nineteenth-Century England, France, and the United States* (Stanford: Stanford University Press, 1981); Margaret H. Darrow, "French Noblewomen and the New Domesticity, 1750–1850," *Feminist Studies* 5 (1979): 41–65; Patricia Branca, *Silent Sisterhood: Middle-Class Women in the Victorian Home* (Pittsburgh: Carnegie-Mellon University Press, 1975).

women as consumers disturbed the separation of spheres through their influence upon the "male" realms of production and capital accumulation. This chapter analyzes mid-nineteenth-century French thought and practice regarding feminine taste and consumption, the relations between "consuming" women and "producing" men, and the implications of consumption as a public function for women that put them in direct contact with men and male activities.

THE FEMINIZATION OF CONSUMPTION

In preindustrial times and during early industrialization women commonly assumed an active role in producing goods and income for the family. They frequently manufactured in the home articles for family use and consumption, and they aided male relatives in small industrial and commercial enterprises, even assuming full responsibility for their operation on the death of a husband.[4] However, the increasing scale of enterprise, the expanding opportunities for educated bourgeois men in public and private administration, and the burgeoning wealth of bourgeois families rendered bourgeois women's productive labor unnecessary and ultimately undesirable. The availability of manufactured products in urban environments reduced the need for women themselves to produce goods at home. The separation of the workplace from the home made it increasingly difficult for women to combine productive with reproductive labor. Finally, a new idealization of family life in the home emphasized interior adornment and comfort, for which someone had to assume responsibility. Since men, with their training and education, were more successful than women at earning income to support a family, the task of home furnishing and maintenance generally fell to women.

Both men and women of the French bourgeoisie explicitly advocated a gender division of function that placed the exercise of taste and the practice of consumption in the realm of women. Constance Aubert, writing for *L'Illustration,* made a very fine distinction when she declared: "Man only needs to know how to spend his fortune; it is only woman . . . who knows how to make of that fortune an

4. Louise A. Tilly and Joan W. Scott, *Women, Work, and Family* (New York: Holt, Rinehart and Winston, 1978); Smith, *Ladies.* This was also true in England. Leonore Davidoff and Catherine Hall, *Family Fortunes: Men and Women of the English Middle Class, 1780–1850* (Chicago: University of Chicago Press, 1987).

entertainment and a glory."[5] What she meant by this was that the
primary role of women was to purchase home furnishings for the
benefit of family comfort, harmony, and even social advancement.
"The great concern of woman is THE HOUSE; the house confided to her
supervision, that she must adorn, make pleasing and accommodating
to the husband who made her its mistress. . . . Mistress of the abode,
she must carefully study all the secrets that make the home agreeable
to the family."[6]

Bourgeois women had considerable domestic responsibilities, including managing servants, caring for children, planning and over-
seeing the preparation of meals, supervising housecleaning, and ful-
filling social obligations. Closely associated with these housekeeping
tasks was the tasteful decoration of the home, and therefore the
discriminating purchase of home furnishings. Whereas Aubert emphasized the importance of female taste and consumption to please
the husband and family, other women writers noted the personal
satisfaction women could gain from the furniture they bought. Constrained—in popular belief and often in fact—to spending most of
their time in the home, bourgeois women not surprisingly treated
pieces of furniture like friends. In an article entitled "My Piano, My
Desk, and My Dresser," Emma Faucon wrote affectionately of these
articles of furnishing, even addressing her desk as she would an
intimate friend. "How many sheets of paper I have blackened on your
green baize! How you would laugh, patient piece of furniture, if you
could express your thoughts, remembering the compositions or essays . . . that have seen the light of day since your arrival in my
room."[7] Similarly, Madame Lafarge recalled in 1841 buying a piano
as part of the marriage gift (*la corbeille*) from her husband-to-be:
"When my choice settled upon a delightful upright piano, I ordered
it to be transported immediately, so that my new friend would arrive
before me at Glandier [Monsieur Lafarge's home in Limousin]."[8]

It is easy to view these statements as a pathetic testimony to
women's confinement in the home. Faucon, an unmarried writer
living with her father and brother, found privacy and solace in her

5. *L'Illustration*, 18 January 1851, 46.
6. Ibid.
7. *Le Conseiller des dames et demoiselles*, May 1860, 206.
8. Madame Lafarge [née Marie Capelle], *Mémoires de Madame Lafarge* (Paris:
Lévy, 1894), 177.

room, among her possessions, after a hectic day of housekeeping. Madame Lafarge, about to begin life with a man she barely knew, in a part of the country unfamiliar to her, clung to an inanimate object as an anchor in a strange sea. But women could also gain more than dreary consolation from household furnishings. Through consumption, they could express something of their personalities and could satisfy a great deal of their ambition for social status and recognition. Moreover, since men often acknowledged women's superior expertise in matters of taste and consumption, tasteful spending could be a means for women to influence men, especially husbands, in the family. Women as consumers actively participated in the formation of the bourgeoisie as a class with tastes distinctive from aristocrats and workers.

Male and female writers of the mid-nineteenth century believed that women had a natural inclination toward tastefulness, simply by virtue of being female—and French. But this inclination still required cultivation in the home and by other women. In her autobiography the writer George Sand (Aurore Dudevant) remembered very clearly her elegant grandmother's efforts to instill in the young Aurore the standards of good taste. "She wanted to form my taste, and trained her discriminating judgment upon any object that attracted me. 'That figure is out of scale,' she would say. 'That combination of colors pains the eye.' 'That composition—those words—that music—that style of dress—is in bad taste.' " Good taste was obviously important to old Madame Dupin, and she may have taken special pains with her granddaughter to counteract the influence of Sand's mother, a woman whose low social origins may have accounted for her own supposedly less discriminating taste. Sand wrote that her mother "had a taste for the new: the latest fashion always seemed the best she'd ever seen. . . . Almost any product of art or industry pleased her, if only it had a beguiling form and fresh colors."[9] Sand never clearly indicated whether, as an adult, she adhered to her grandmother's or to her mother's standard of taste, though she implied greater appreciation for the former's schooled discrimination. Sand, of course, in her unconventional (for a woman) career as a writer, had many occasions to exercise her taste beyond the purchases

9. George Sand, *My Life* (1879), trans. Dan Hofstadter (New York: Harper and Row, 1979), 87–88.

of household furnishings. By contrast, male thinkers conceived of feminine taste as something rigidly limited to home and family.

The moral philosopher Paul Janet wrote some time after the exhibition that only in the home and from women could men acquire certain characteristics that they naturally lacked: "refinement, taste, and the feeling for art which woman projects into everything and wants to see realized or expressed in the objects and beings that surround her."[10] Janet suggested that men could learn about taste from women, but strictly within the domestic context and from the way women exercised taste in interior decoration. Taste gave women an edge over men, who depended upon women's arbitration and teaching in such matters, but only in the privacy of the family. A close contemporary of Janet's, the art critic Léon de Laborde, wrote even more extensively on feminine taste. As part of his report on fine arts at the Crystal Palace exhibition, Laborde argued that cultivating women's inherent tendency toward tastefulness and art was both a social and an economic imperative if France were to maintain its reputation for producing tasteful and beautiful consumer goods: "To teach [art to] a woman is to create a school in the family."[11] To be sure, Laborde also drew up an elaborate plan for formal education in aesthetics and drawing for talented boys and girls, but he still found children's early learning about taste in the home and from a mother to be indispensable to his larger project of an artistically educated population in France.

TASTE AND POWER IN GENDER RELATIONS

While these pronouncements on feminine taste allowed only a limited realm for its exercise, literary and prescriptive works indicated that the role of tasteful consumer for the home could also be a source of power for women in their relations with men. By turning over part of their income to their wives for household expenses, and by conceding to women the task of interior decoration, bourgeois husbands

10. Paul Janet, "La Famille," *Grand Dictionnaire universelle du XIXe siècle* (Paris: Larousse et Boyer, 1866–90) 8:75.
11. Commission française sur l'Industrie des Nations, *Exposition universelle de 1851: Travaux de la Commission française sur l'Industrie des Nations* (Paris: Imprimerie impériale, 1855) 7:523. Also see chapter 6.

occasionally subordinated themselves to their wives' tastes and ambitions.

In Flaubert's *Madame Bovary*, Emma has had a convent education (a sign of refinement in the rustic area around Rouen), and she consults fashion magazines from Paris; she quite overwhelms her boorish husband Charles with her decoration of their home.

> She wanted two large blue vases for her mantelpiece, and some time after an ivory *nécessaire* with a silver gilt thimble. The less Charles understood these refinements the more they seduced him. They added something to the pleasure of the senses and to the comforts of his fireside. It was like a golden dust sanding all along the narrow path of his life.[12]

Emma, though denied the professional training available to men, nonetheless has acquired something of the cultivation her husband lacks—a not uncommon occurrence in bourgeois marriages.[13] Her efforts at tasteful consumption for the home, a manifestation of her social ambition for herself and her family as well as a means of personal gratification, awaken Charles to his own deficiencies in a pleasant and bewitching way. He is more than ever in thrall to Emma as a result of the clothes and furnishings she buys.

In a short story by Louis Veuillot, a young bourgeois husband acknowledges the significant contribution of his wife to his domestic bliss, and ultimately to his professional success. Cléante, a bureaucrat, turns over to his wife, Lucile, the management of household funds, with happy results. "Yesterday [Cléante] did his yearly accounts, and he found that, thanks to the strict administration of Lucile, his household had cost him less since he was married. . . . He congratulates himself on having chosen his wife. . . . She reigns by grace, by richness, by virtue."[14] Obviously Cléante was pleased that Lucile saved him money; but the terms "grace," "richness," and "virtue" refer to her impeccable taste in matters of dress, furnishing, and behavior. Because of Lucile's good taste and capable management, Cléante makes a good impression on his friends and associates,

12. Gustave Flaubert, *Madame Bovary* (1857), trans. Eleanor Marx-Aveling (New York: Jonathan Cape and Harrison Smith, 1930), 74.

13. Adeline Daumard, *La Bourgeoisie parisienne de 1815 à 1848* (Paris: SEVPEN, 1963), 362–63, 366.

14. L. Veuillot, "L'Honnête Femme," *Le Correspondant* 1 (1843): 234.

and Lucile is confident that he will rise quickly in the bureaucratic hierarchy. This is all to the good for Cléante; but the author of the story also notes that "Lucile played her husband like an organist plays his instrument, and this art is not rare among women."[15]

What Pierre Bourdieu calls "cultural capital"—that is, the education and especially the cultivation of good taste that women bring to a marriage—was so important in nineteenth-century France that several writers maintained that a bourgeois marriage would fail without it.[16] In a story about the 1855 exhibition, Jenny de Fresne describes a young man observing the women who pass in front of the exhibits. He watches many females stop and admire displays of lace and other feminine finery, but the woman who attracts his attention as a potential wife is the one who stands before housewares. De Fresne acknowledges the wisdom in the young man's method of selecting a bride who would probably be a good housekeeper rather than a frivolous spendthrift, but she cautions that not only homemaking skills but also good taste are indispensable accomplishments of the ideal wife.

> A good homemaker is undoubtedly a precious treasure; but if she is no more than that, if her parents deprived her of all intellectual culture, if she has no feeling for the arts, she might make the joy and happiness of an honest worker, but she destines for the man of the world—a destiny she will share—the most bitter disenchantments. Happy is he who encounters in his companion the gifts of mind and heart, united with the inestimable qualities of the housewife![17]

The message here is that a woman who did not please her more worldly husband through her artistic sensibility and home decoration risked an unhappy marriage. Indeed this is precisely what happens to poor Augustine in Balzac's story "La Maison du chat-qui-pelote." Augustine is the daughter of a very successful draper, and one sign of the tradesman's social rise is the marriage of Augustine to the aristocratic painter Sommervieux. Not long after the wedding Sommervieux increasingly spends his time away from home, visiting the

15. Ibid., 103.

16. Pierre Bourdieu, *La Distinction: Critique sociale du jugement* (Paris: Éditions de minuit, 1979).

17. *Le Conseiller des dames et demoiselles,* September 1855, 326.

duchess of Carigliano, despite Augustine's best efforts to induce him to remain in her company. She finally decides to see for herself what the attraction is for her husband at the duchess's apartment. Though untutored in the ways of refined living, Augustine discerns immediately on her arrival at the duchess's Saint-Germain address that the aristocratic woman's interior decoration is more enticing to a man than Augustine's natural beauty.

> As she made her way through the stately corridors, the handsome staircases, the vast drawing-rooms . . . decorated with the taste peculiar to women born to opulence or to the elegant habits of the aristocracy, Augustine felt a terrible clutch at her heart; she coveted the secrets of an elegance of which she had never had an idea; she breathed an air of grandeur which explained the attraction of the house for her husband.[18]

The decoration of her rooms is an extension of the duchess's personality, to such an extent that Augustine thinks in terms of the *house's* attraction for her husband, not the duchess's. Indeed, even before meeting the duchess, Augustine concludes, solely on the basis of the exquisitely, luxuriously decorated apartment, that "as a woman, the duchess [is] a superior person."[19] The good taste of the aristocracy in general provided a model for female bourgeois consumers, but in this case the newly rich and uncultivated Augustine cannot compete with a woman born to wealth and surrounded by elegance. Augustine's deficiency as a tasteful decorator of the home is also a failure in female attractiveness and perhaps in femininity itself.

Women were thus seen as seducing men through interior decoration, through the good taste and careful consumption that men were too busy to acquire or practice.[20] Female identity within the

18. Honoré de Balzac, "La Maison du chat-qui-pelote," in vol. 1 of *Oeuvres complètes de M. de Balzac* (Paris: Les Bibliophiles de l'originale, 1965), 76.

19. Ibid.

20. Another example in literature of a woman seducing a man through the decoration of the home appears in Zola's *La Curée* (1871). In this instance a fashionable conservatory, warm with steam and filled with exotic plants, turns Renée, the bored and passive wife of the speculator Aristide Saccard, into the predatory seducer of her stepson, Maxime. Zola, of course, saw nothing tasteful in the extravagant furnishings of the new rich (such as the Saccards) under the Second Empire, but he acknowledged the influence of domestic interiors on gender relations. Émile Zola, *La Curée* (Paris: Fasquelle, 1984), 204–9.

bourgeoisie was so closely tied to tasteful consumption that a woman who failed in this endeavor risked losing not only her husband but also her femininity, a fact that both Sand and Balzac noted in connection with Sand's experience. When Sand left her husband to live in Paris on her own, she found life very difficult on her small allowance of 1,500 francs. "The hardest thing was to buy furniture. Elegance was out of the question, of course." Though Sand did indeed manage to furnish her fifth-floor apartment, arrange for her meals to be brought up from a restaurant, and engage her concierge to help with the cleaning, she recalled Balzac's incisive observation: "'You can't be a woman in Paris on under twenty-five thousand [francs].' And this paradox, that a woman was not really a woman unless she was smartly dressed, became a reality for the woman who would be an artist."[21] Sand claimed that she resorted to wearing men's clothes because she could not afford the expense of conventional female dress. Whether or not this was the only or the true explanation for Sand's notorious cross-dressing, the point about money, consumption, and femininity is important. Sand was on the margin of respectable bourgeois society, not only because she left her husband and embarked on a career but also because she could not consume in a manner appropriate to her social position and her sex. She voluntarily abandoned the superficial signs of bourgeois femininity by separating from her husband and sacrificing the marital income that had allowed her to dress well and live elegantly. For it was generally a husband's income or a joint, marital income that permitted women to consume for home and family. Men thus contributed financially to women's cultural ascendancy in the home. In addition, the very control of money for household expenses constituted another aspect of the power women as consumers wielded in the family.

According to the feminine press, women's control over the household budget could greatly affect the financial solvency and the social status of the family. "It is by the active intelligence of the homemaker, by her attentive and well-directed supervision, that modest fortunes are augmented and large fortunes preserved."[22] However, the capacity of women to conserve family wealth through sensible consumption should not, according to women's periodicals, go so far as to become niggardliness. On the contrary: "It must be understood

21. Sand, *My Life,* 203.
22. *Le Foyer domestique,* 1 June 1850, 418.

that domestic economy means not the reduction but the intelligent and fruitful distribution of the household budget. The mistress of the home, like the finance minister of a great state, has capital to apportion."[23] "Fruitful" is the key word here. Women, like nations, had to keep up appearances for reasons of status and power. A story in a woman's periodical presents a happily married couple, who nonetheless disagree about household spending. "Madame occasionally accuses Monsieur of being too tight with money, of being too economical, their position putting them, she says, above such trifling; and even more [she accuses him] of following too much the paths trodden by the vulgar, in a word, of being too much like everyone else."[24] The wife's twitting in this case leads the husband to loosen his purse strings and purchase a piano that his wife deems necessary to their mutual happiness and social standing. Husband and wife are both sensitive to the importance of appearances based on consumption, but it is the woman's role and duty to determine the occasions and the extent of spending on household furnishing and clothes.

This condition is the object of caricature in a story of 1846, in which a bourgeois woman scrimps to an extreme on family expenses — daily meals, household lighting — in order to buy an expensive cashmere shawl, a status symbol for women of her class. The bemused husband hardly knows what is going on, except that his wife is being admirably economical; and he is afraid to indulge in his habit of buying an occasional book. But the story concludes that the shawl represents only the beginning of the wife's desire for improved status by means of material possessions; her next passions will be finer clothes for herself to match the quality of the shawl, followed by better clothes for her husband. "Then she will find the furniture not up to par with the cashmere shawl; she will redo the upholstery like a new outfit; after the furniture, the apartment."[25] The story represents a nightmare for husbands, victims of their wives' consuming ambition, and unable to escape from this exaggerated fulfillment of womanly duty. Underlying this unflattering portrait of the female consumer, however, was the fact that women's status did rest to a considerable extent upon their access to money for household consumption.

23. *Le Conseiller des dames,* December 1847/January 1848, 98.
24. *Le Foyer domestique,* February 1854, 131.
25. *L'Illustration,* 7 November 1846, 151.

THE FRENCH BOURGEOIS WOMAN AS CONSUMER

In support of Audiganne's contention that women in England exerted less influence as consumers than in France, Flora Tristan, the social critic, deplored the low status of English women because their husbands, rather than they, controlled spending for the home. This recognition came from Tristan's observation during her travels in London that "the woman in England is not always, as in France, the mistress of the home. . . . The husband keeps the money and the keys; it is he who regulates spending."[26] According to Tristan, for lack of access to money English women were lamentably subordinate to men, compared to their French sisters. Status and influence for women of the bourgeoisie depended upon control over money and the cultivation of taste.

To what extent can we trust fictional and journalistic sources in determining the actual influence of feminine taste upon men and the family, and the extent to which women controlled financial resources? They are almost the only sources on the elusive matter of taste and the private matter of family spending. Letters and memoirs of bourgeois women occasionally mention mundane activities like shopping or decorating, but in general the women who wrote their memoirs for publication focused on the more extraordinary aspects of their lives—acquaintances with the famous, literary successes, emotional traumas, or physical crises. A carton full of household account books in the National Archives in Paris indicates that at least some women had an allowance for household necessities and small personal expenditures on items like food, sewing notions, cleaning materials, transportation costs, and furniture and clothing repair.[27] But these do not reveal who spent what on higher-priced articles of furnishing. A

26. *La Gazette des femmes,* 11 February 1843. This passage first appeared in Tristan's *Promenades dans Londres,* which has recently been translated into English as *Flora Tristan's London Journal, 1840,* trans. Dennis Palmer and Giselle Pincetl (Boston: Charles River Books, 1980), 195–97. To some extent Branca's research on English women of the Victorian era confirms Tristan's observation, though the causality is questionable. Branca writes: "Because the middle-class woman was given little responsibility for administering or organizing in the family, she was often considered untrustworthy." Branca, *Silent Sisterhood,* 8.

27. Archives Nationales, Fonds privés, AB XIX 3503, Comptes de ménagères, 1878–1900. See also 246 AP 40, Papiers de Mme H. Fortoul, 1849–1856; and *Comptes d'un budget parisien: Toilette et mobilier d'une élégante de 1869* (Paris: F. Henry, 1870).

set of receipts kept by a journalist and her husband revealed that the husband was almost always billed for furniture, works of art, glassware, and china; but it is impossible to know if this was merely convention—by law husbands exercised great control over the couple's goods in marriages contracted under the principle of community property—or if he actually walked into those stores and made the purchases himself or with his wife.[28] Marriage contracts themselves indicate that both husbands and wives brought some of their own clothing and furniture to the new household, but inventories at the time of death of a spouse suggest that many more purchases occurred at the beginning of and during the marriage.[29] For instance, living-room sets of sofas, chairs, and armchairs all made of the same wood and upholstered with the same fabric were probably a joint purchase at the time of or shortly after marriage.

Both autobiographical and prescriptive literature support the finding that marriage was often the occasion for young women to exercise their taste as consumers in furnishing a home. Madame Lafarge recalled that when she was a bride she had free rein to decorate her husband's home as she wished: "[Madame Lafarge senior] told me that I would be absolute mistress [of the house], and that I could act like a despot and alter my new abode according to my tastes and my habits."[30] And in fact Monsieur Lafarge encouraged his young wife to elaborate her detailed plans for major renovation of the family home. Later in the century young Caroline Brame was positively smitten with the way her newly married cousin decorated her apartment:

What a pretty house! How well it is decorated. The salon is especially delightful. The curtains are of sky-blue satin, the furniture of the same, with Louis XV medallions; the small console on which these many graceful nothings are scattered—all of this is fresh, delightful! Everything is chosen with taste.[31]

28. Archives Nationales, Fonds privés, AB XIX 3503: K-M 106, Comptes de Mme Parizet à Taverny, 1884–1886; and S 106, Lot de factures anciennes, 1878–1900.

29. Daumard, *La Bourgeoisie,* 136; Archives Nationales, Minutier Central, Inventaires après décès, XVIV 917 to 919 (1830); XXIX 1128 to 1131, 1133 (1850). See chapter 3 for complete citations of the death inventories consulted for this study.

30. Lafarge, *Mémoires,* 202.

31. *Le Journal intime de Caroline B.,* ed. Michelle Perrot and Georges Ribeill (Paris: Montalba, 1985), 46.

Brame's cousin and Madame Lafarge were only following the guidelines of the feminine press, with its frequent articles on what kinds of
items a young bride should purchase for furnishing her new household.

From the available sources it is clear that tasteful consumption was
an essential feature of bourgeois women's lives. Ideally, girls learned
about taste from their mothers, or other female relatives, at an early
age; and this expertise made them attractive to young men as future
wives. Tasteful consumption carried different connotations for different women, depending on their characters and their marital and
social situations. For some, buying clothes and furniture might be
one of the few socially acceptable activities through which they could
express their personalities and fulfill their desires. For others, however, consumption could be an assertion of female ambition for
personal and family status, and a counterweight to male authority in
the home. Women could seduce men, advance them, even break
them with their spending for the household. To be sure, fictional
representations of women and their consumption exaggerated the
reaches of female spending and their effect on men. Nonetheless, on
a more modest scale, women of the bourgeoisie could indeed influence a husband's career or a family's social position through control
of household finances and the kind of taste exercised in spending.
Men and women of the bourgeoisie participated in a division of labor
whereby males earned income and females spent it; the relationship
between earner and spender could be both complementary and conflictual. Up to now, scholars have emphasized the complementary
features of this form of social organization. The first to do so was
Thorstein Veblen in his study of conspicuous consumption in the
late-nineteenth-century United States.

THE WIDER IMPLICATIONS OF FEMALE CONSUMPTION

In his *Theory of the Leisure Class,* published in 1899, Veblen contended
that among the new rich of the United States, the women's conspicuous consumption was a symbol of leisure-class status. Neither the
men nor the women of this new social and economic elite were born
to leisure and wealth; and the men took pride in the fact that their
women were idle and could spend their time purchasing frivolous and
luxurious articles of clothing and furnishing, thanks to their men's

success at making money through industry, trade, or finance. Women's ostentatious display of wealth through extravagant spending complemented men's hard work at accumulating wealth.[32] Veblen's analysis, however, ignores the potential for conflict in this particular separation of spheres, both at the private, domestic level and at the public level of social production and reproduction.[33] The examples discussed earlier in this chapter show that women's consumption could go beyond compliant accommodation of their husbands' and families' desires for domestic comfort. Women's tastes for expensive symbols of social status could come into conflict with men's valuation of rational accumulation; and since both parties were acting within the conventional boundaries of female and male behavior, the ensuing struggle might be fairly evenly matched.

Beyond the home, women's demand for tasteful consumer goods constrained male producers to retain methods of manufacturing that were not always as efficient as available technology and control of capital allowed. To be sure, manufacturers adopted division and dispersal of labor, among other strategies, to meet the increasing demand for consumer goods that were artistic, unique, and novel.[34] But still more efficient methods, like mechanization and concentration of manufacturing, were often poorly suited to producing the kinds of goods that bourgeois women sought, as chapters 4 and 5 will show. The point here is to explore consumption/production relationships beyond the demographic explanations of economic historians.[35]

32. Veblen, *Leisure Class,* 80–85.

33. More recent theorists of consumption similarly acknowledge the gendering of consumption as feminine, but fail to analyze it. Bourdieu, *La Distinction;* Edmond Goblot, *La Barrière et le niveau* (1925; Paris: Presses universitaires de France, 1967); Daniel Miller, *Material Culture and Mass Consumption* (Oxford and New York: Basil Blackwell, 1987).

34. Alain Faure, "Petit Atelier et modernisme économique: La Production en miettes au XIXe siècle," *Histoire, économie et société* 4 (1986): 531–57; Ronald Aminzade, "Reinterpreting Capitalist Industrialization: A Study of Nineteenth-Century France," in *Work in France: Representations, Meaning, Organization, and Practice,* ed. Steven Laurence Kaplan and Cynthia J. Koepp (Ithaca, N.Y.: Cornell University Press, 1986), 393–418.

35. Patrick O'Brien and Caglar Keyder, *Economic Growth in Britain and France, 1780–1914: Two Paths to the Twentieth Century* (London: George Allen and Unwin, 1978); Maurice Lévy-Leboyer, "Les Processus d'industrialisation: Le Cas de l'Angleterre et de la France," *Revue historique* 239 (1968): 281–98; Rondo E. Cameron, "Economic Growth and Stagnation in France, 1815–1914," in *The*

The extreme gendering of consumption and production, notably in bourgeois families, might also have contributed to the slow pace of growth in manufacturing efficiency in nineteenth-century France. Oriented from a young age to eschew common, uniform, flashy articles of clothing and furnishing, bourgeois women rejected in consumer goods the values that promoted efficiency and financial gain for male capitalist producers. This is not to deny that many bourgeois entrepreneurs made a great deal of money from their putting-out establishments and their factories, or that modern methods of production were gradually adopted throughout France during the nineteenth and twentieth centuries. But gender divisions within society were neither as neat nor as complementary as their proponents chose to assume. The interests and values of female bourgeois consumers were not necessarily the same as those of male bourgeois producers. Nor were these two groups as separated from one another as the "separate spheres" notion might suggest.

The idea of separate spheres, as Bonnie Smith analyzed it with regard to bourgeois women of the Nord region of France, meant that females concocted for themselves a "culture of the home" based on values and activities entirely different from those of men, who occupied the public realm of business and politics.[36] Smith's insightful analysis of the female world ignores the actual mingling of males and females that occurred in family and social situations, nor does it account for the interaction of men and women (and the intersection of male and female spheres) when women fulfilled their feminine role of consumer. What happened when women, schooled in the bourgeois standard of taste for domestic comfort and pleasure, walked out of their plush and padded homes and into the streets and shops of the city to negotiate with shopkeepers and producers? For adherents of the separation of spheres, this ordinary activity represented a puzzling breakdown of barriers between the female and the male space, even as women and men carried out their proper, gender-specific functions.

A sketch that appeared in a fashion magazine suggested that when women shopped they assumed a different, "unfeminine" character and behavior.

Experience of Economic Growth: Case Studies in Economic History, ed. Barry E. Supple (New York: Random House, 1963), 328–39.

36. Smith, *Ladies.*

With regard to shops women are essentially idlers, artists, and fond of all that glitters, of all that blazes in the rays of natural sunlight or of the sunlight of gas jets. A woman, whatever her age, intrepidly crosses from one side of the boulevard to the other, braving the messiest pavement, the dust, the streetcar, in order not to miss a store that attracts her as the mirror attracts the lark, the brilliant reflection of a brocaded satin, or the silky and gentle gleam of a watered silk.[37]

Here the female shopper was no longer a subservient, dutiful, confined domestic creature. She tenaciously pursued her quest for shining treasure. Was this single-minded, determined, and intrepid shopper the same person, the same sex, as the dutiful housekeeper, attentive to the needs and desires of her family? Certainly she was, but the conventions and stereotypes of feminine character and behavior did not easily accommodate this contradiction within the female ideal. On the one hand, the gender division of labor allocated household consumption to women as an essential complement to the production, commerce, and administration that constituted the male realm. On the other hand, consumption sanctioned a public role for women outside of the home, one that in the second half of the century department stores would increasingly foster and exploit. The Crystal Palace exhibition underscored this contradiction between the private ideal and the public fact of female consumption by welcoming women as potential buyers of the kinds of goods on display. In a sense the exhibition was a prototype department store with its wide range of consumer goods; from this perspective it was only right that women as consumers, as well as men as producers, should view the exhibits. But the political economist Adolphe Blanqui's view of women in the Crystal Palace reveals some confusion, even dismay, over their presence in what men usually referred to as an exhibition of industry, not of consumption.

CONTRADICTIONS IN THE SEPARATE SPHERES IDEOLOGY

In a short aside on the numbers of women who visited the Crystal Palace exhibition, Blanqui suggested that their appearance in a public (and therefore male) place that exalted production was both monstrous and at the same time a fulfillment of feminine obligations.

37. *Le Moniteur de la mode*, January 1861, 390.

Women are in the majority here, and one would think that the English [men] organized the exhibition for the women out of pure gallantry. They are tireless. They eat like ogres at all the refreshment stands. The detestable fashion of crinolines and even of panniers . . . gives them a really fantastic volume that daily reduces the free space left for circulation. Our unhappy stars must try hard not to get caught in the orbit of these immense planets that crowd like distant suns, cold and unknown to astronomy, in the world of the exhibition. There is even something strange and odd in seeing this exhibition within the exhibition; but it proves at least that women here take . . . a real role in the progress of industry and that they seriously concern themselves with the interests and work of their husbands.[38]

Blanqui found the female spectators at the exhibition "strange" and "odd"; he compared them to ogres. From his perspective, women shoved men aside with their huge skirts, voraciously consumed the available food, and wore out their male companions with their energy. In his astronomy metaphor Blanqui equated these women with cold, distant, unknown planets or suns, oblivious to the inconsequential men trying to navigate around them. Clearly Blanqui did not know quite what to make of the women in the Crystal Palace. In an exhibition of industry that Blanqui considered to be staged by and primarily for men, women were foreign bodies.

Blanqui's confusion about the presence of women in the Crystal Palace was also expressed, rather awkwardly, in some caricatures by Cham (Amédée Charles Henri Noé), for the magazine *Charivari*. The artist drew a series of comic sketches about the exhibition, most of them representing imagined, extraordinary inventions that went wrong in some fashion. However, women were central in some instances, and Cham portrayed them both as objects on display for male admiration and as hindrances to men's enjoyment. For example, in one cartoon a man approaches an exhibit in the Crystal Palace as an excuse to get close to a pretty female visitor (fig. 7). By contrast, in another cartoon the exhibition served as a backdrop for marital conflict; a husband is about to throw his wife into the Crystal Fountain to avoid having to take her to a watering spa (fig. 8). Much more ambiguous in terms of gender relations is the cartoon of a man

38. Adolphe Blanqui, *Lettres sur l'Exposition universelle de Londres* (Paris: Capelle, 1851), 89–90.

sitting on a woman's lap outside the Crystal Palace (fig. 9). The cartoon's caption—"On leaving the exhibition one feels an urgent need to sit down anywhere"—both supports Blanqui's claim that women took up too much space and reverses it by showing a man undeterred in the pursuit of his object by feminine precedence. Whereas Cham's intention was merely to give readers a laugh, Blanqui seemed compelled to explain the appearance of women in the Crystal Palace more realistically. He tried to ease his discomfort with their overwhelming presence by attributing that presence to women's interest in their husbands' affairs.

To be sure, Blanqui appreciated the role of the female as consumer, and elsewhere in his account of the exhibition he accepted with pride women's homage to the manufacturing superiority of France: "One sees . . . English ladies in ecstasy before our display of shawls."[39] But Blanqui never allowed women of the bourgeoisie to get too close to bourgeois men's realm of production. When he wrote of French women as consumers, he associated them with female producers in such a way as to deplore a system that permitted, even encouraged, wage-earning women:

> Women among the fortunate of the earth, I cannot tell you too often: when you toss over your beautiful shoulders these airy [lace] scarves, think sometimes of the poor girls who made them. They are of your sex, of your country, and of your religion, and they often lack basic necessities, after providing you with superfluities![40]

Though Blanqui exhorted female consumers to think of their less fortunate sisters who had to labor for a living, he offered no specific remedies for the situation, not even charity. He simply lamented the fact that some women were poor. Moreover, by associating female consumers with only female producers, Blanqui awkwardly tried to maintain the fictional separation of spheres that kept women and men apart. Obviously women who shopped (or who went to the exhibition) and women who worked left the home temporarily and entered the public (male) sphere. Blanqui could accept this departure from domesticity on the part of female consumers because they were, in a way, still being domestic. But wage-earning women were another

39. Ibid., 50.
40. Ibid., 132.

matter; Blanqui would have preferred to see them caring for their own families instead of working for wages, even if the wages were intended for family use.[41] In Blanqui's perfectly ordered world, women of both classes would be separated from male producers and under their sway in the home, an ideal that he tried to convey in emphasizing the sexual similarity above the class difference when he addressed female consumption.[42]

CONCLUSION

The gender separation of spheres was a convenient ordering of society that privileged male productive and political functions, but such a division was not and could not be perfect. Men and women shared public and private spaces in both complementary and conflictual relations. The only way to understand the social, economic, and political structures that the bourgeoisie dominated in nineteenth-century France is to account for the contributions of both men and women to this organization.

The evidence presented here suggests that bourgeois women in France took seriously their role as taste makers and consumers for the family. In this way they not only reproduced and reinforced social hierarchies and bourgeois values but also enhanced the strength of their position in the family and with regard to men. The Crystal Palace exhibition revealed yet another problematic aspect of designating consumption as both a feminine and a domestic task: by welcoming women as spectators and even, indirectly, as judges of the goods on display, the exhibition revealed that women had to venture into the public space of the marketplace, where they influenced the "male" realm of production through their demand for unique, tasteful, and handmade goods. Thus women as consumers played a significant social and economic role in mid-nineteenth-century France, a role perhaps larger than contemporaries accorded them. The display of industry at the Crystal Palace exhibition has hitherto

41. Adolphe Blanqui, *Des Classes ouvrières en France pendant l'année 1848* (Paris: Firmin Didot frères, 1849), 211–12.

42. Joan W. Scott, "Emergence of the Problem of the Working Mother" (Paper presented at the Seventh Berkshire Conference on the History of Women, Wellesley, Mass., 19–21 June 1987); Joan W. Scott, "Statistical Representations of Work: The Politics of the Chamber of Commerce's *Statistique de l'Industrie à Paris, 1847–48*," in *Work in France,* ed. Kaplan and Koepp, 335–63.

been analyzed explicitly or implicitly as a male event; but the social structures and economic relations underlying the manufacture of consumer goods required the contributions of both men and women, in their respective gender roles and outside of them. Indeed the ideal of an exclusively private or domestic role for bourgeois women breaks down in the practical activity of consumption; and this breakdown is reflected in male ambivalence toward female shoppers and consumers in public.

Analysis of the tastes and habits of consumers in mid–nineteenth-century France elicits a complex of social and gender relations essential to bourgeois existence and identity. But how did these values and practices translate into actual purchases—into the material life of the bourgeoisie? Addressing this question requires leaving the public space of the Crystal Palace exhibition and peering into the homes of the middle class to see what consumers bought and to determine the interaction between consumption and production hinted at in accounts of the exhibition.

Chapter Three

Symbols of Status, Signs of Change

Furnishings in the Bourgeois Household

The preceding chapters have emphasized the quest for symbols of ruling-class status and the evolution of a domestic role for women; but there were other considerations that affected consumer choices. Innovations in raw materials, division of labor, and mechanization brought new products onto the market, modified existing products, and usually lowered prices. At the Crystal Palace exhibition iron beds, papier-mâché furniture, zinc statues, and rubber raincoats demonstrated to visitors that familiar objects could be made of new materials. Fans, toys, carved wooden furniture, and chased and sculpted silver tea sets showed potential consumers what division of labor in production could accomplish; such items required as many as twenty workers each performing separate tasks in the completion of the product. Finally, machine-woven carpets and shawls, electrolytically silver-plated tableware, and lithograph prints revealed the capacity of new technologies to imitate expensive goods at prices that middle-class consumers could afford. The bourgeoisie, sensitive to price and attentive to novelty, had to weigh these factors against a desire to appear rich, powerful, and discriminating according to standards adapted from the previous ruling class, the aristocracy.

What, then, did bourgeois consumers buy? Did they follow the advice of the press and furnish their homes with artistic, stylish, handmade goods? Did they take advantage of new opportunities for cheaper, novel items that emphasized comfort and convenience above status? Data on bourgeois households in Paris, both before and after the Crystal Palace exhibition, provide a basis for understanding the interaction between consumer taste and the technological capabilities of French industry. By comparing the numbers and kinds of household possessions over time, it is possible to locate changes in consumption practices in the larger context of long-term developments in social and economic conditions and gender relations.

Bourgeois consumers were fairly uniform in what they regarded as essential articles of furnishing for members of their class. In a small sample of bourgeois households in Paris during the Restoration and the Second Empire, beds, chairs, tables, dishes, cookware, and heating and lighting devices were ubiquitous. Even the form and arrangement of these items were remarkably similar within each period. Moreover, the repeated appearance of certain decorative or nonessential objects in the sample strongly suggests a consensus across various income levels and occupations on what it meant to be bourgeois. But there were also changes over time in both the basic and the decorative articles of household furnishing; these changes were linked to technological, economic, and social developments. Tracing the appearance, absence, and modifications of particular furnishings from the Restoration to the Second Empire indicates that a consumer desire for status—manifested in the purchasing of items that reflected aristocratic tastes and habits from the past—remained prevalent among Parisian bourgeois. Yet in certain cases where comfort, convenience, and low cost did not directly challenge the widespread consensus on taste, consumers adopted new or differently made articles of furnishing. Other factors, such as gender relations, fear of social unrest, the rebuilding of Paris, and the general rise in bourgeois fortunes, also affected consumer tastes. The following description and analysis of Parisian bourgeois' possessions assert the importance of consumer practices for understanding the nineteenth-century French bourgeoisie.

THE SAMPLING OF HOUSEHOLD INVENTORIES

Inventories of possessions made at the time of the death of a householder (*inventaires après décès*) are the basis for this chapter's analysis of bourgeois taste and consumer practices.[1] These inventories are part of the notarial records (Minutier Central) of the Archives Nationales;

1. The inventories that comprise the sample are from the Archives Nationales, Minutier Central: VI 1230 (23 February 1870); XII 1224 (5 January 1870); XXXI 601 (4 March 1830); XXXV 1098 (1 October 1829), 1100 (24 April 1830); XLI 1131 (8 and 27 April 1870), 1134 (4 and 9 July 1870); XLV 954 (9 February

they consist of detailed listings of household furnishings and their approximate value.[2] A major objective in consulting the inventories was to determine to what extent bourgeois consumers actually adhered to the dictates of good taste put forward by the feminine press and in accounts of the Crystal Palace exhibition, by purchasing artistic, elegant, handmade goods. Additionally, I wished to learn to what extent the technological changes heralded at the exhibition altered ordinary people's lives, by introducing new products into the home or modifying familiar goods to increase comfort, enhance elegance, lower the price, or promote convenience. For this purpose I selected sixteen inventories from the Restoration (1815–30) and sixteen from the Second Empire (1852–70). The small size of the sample is due to the considerable difficulty of locating appropriate inventories among the masses of notarial documents, which are classified by notary and date rather than by type of document. Nonetheless, even this small sample allows some conclusions to be drawn about consumer trends, and other primary and secondary sources support those conclusions.

The time span separating the two periods was sufficient to allow for visible changes in consumer practices; yet the two periods are united by the continuous rise of the bourgeoisie to economic, political, and social dominance, finally consolidated under the Third Republic (1870–1940). This time frame makes it possible to see which developments highlighted at the exhibition were reflected in the lives of Second Empire consumers but not in those of their Restoration counterparts.

The inventories were all taken in 1828–30 or 1869–70. The sample includes only those households that were formed (that is, the husband

1870), 958 (24 November 1870); LVI 648 (19 August 1828), 652 (6 March 1829), 657 (11 September 1829), 663 (15 June 1830), 664 (31 August 1830), 666 (6 November and 17 December 1830); LX 910 (1 February 1870), 911 (14 and 30 March 1870); LXXV 1338 (30 April 1870), 1340 (29 June 1870); LXXVI 946 (13 December 1869), 950 (10 August 1870); LXXVII 587 (3 February 1830), 588 (7 April 1830), 590 (22 December 1830); LXXIX 600 (18 February 1830), 607 (24 February 1830), 609 (4 May 1830); CVII 963 (20 January 1870).

2. Though other historians have consulted these documents for illuminating the material life and culture of ordinary people in France, none have used them to address the subject of bourgeois consumerism and its relationship to industrial development. Daniel Roche, *The People of Paris,* trans. Marie Evans with Gwynne Lewis (Berkeley: University of California Press, 1987); Adeline Daumard, *La Bourgeoisie parisienne de 1815 à 1848* (Paris: SEVPEN, 1963); Pierre Sorlin, *La Société française,* vol. 1, *1840–1914* (Paris: Arthaud, 1969).

and wife were married) during the Restoration or the Second Empire, respectively. Most couples furnished their homes at the time of marriage with newly purchased goods.[3] Thus, for an inventory to reflect reliably the consumer tastes of the Restoration or of the Second Empire, the household should have come into existence during the same regime. Only intact nuclear families, with or without children, were included in the sample (no widowed or single persons), and only couples married under a community-property agreement. Other criteria for inclusion were a "bourgeois" occupation for the head of the household (shopkeeper, manufacturer, merchant, bureaucrat, or professional), and a total value of household goods amounting to at least 3,000 francs. This figure is somewhat arbitrary. The dividing line between petty bourgeois and working-class status was notoriously blurry, but households with possessions valued at 3,000 francs had features not characteristic of working-class habitations, such as several rooms, ample furnishings, an array of decorative objects, and a servant.[4] No upper limit was adopted for the value of inventories, but households of titled persons were excluded. Judging from a few inventories of aristocratic households, they were distinguished from bourgeois homes by having more rooms, more expensive furnishings, works of art, horses and carriages, and more servants.

A strength of the sample is its wide range of bourgeois households, from petty shopkeeping families crowded into small quarters alongside of or above the business, to successful professionals and merchants living with their well-dressed wives and children in elegantly furnished and spacious apartments. Within this range, the inventories suggest that all shared a similar level of material comfort and household practices. A description of the Gruyer household during the Restoration provides an orientation to typical bourgeois furnishings.

BOURGEOIS HOME LIFE IN THE RESTORATION

Marguerite Adèle Vaude and Nicolas Antoine Gruyer were married in 1815. In the following eleven years, Madame Gruyer bore a daughter and two sons. Monsieur Gruyer was the owner of a carpentry

3. Daumard, *Bourgeoisie,* 136. See also chapter 2.
4. Michelle Perrot et al., *De la Révolution à la Grande Guerre,* vol. 4 of *Histoire de la vie privée,* ed. Philippe Ariès and Georges Duby (Paris: Éditions Seuil, 1987), 314–19, 361; Frédéric Le Play, *Les Ouvriers européens,* 2d ed. (Paris: E. Dentu, 1878) 6:327–492.

establishment, and at the time of his death in 1829 the family lived
just north of the chic Chaussée d'Antin area on the rue Blanche, next
to and above the carpentry shop.[5]

Monsieur and Madame Gruyer must have enjoyed a good table, if
not on a daily basis, then at least on holidays and other festive
occasions. In addition to the usual supply of ordinary Bordeaux table
wine, the Gruyer wine cellar was stocked with several bottles of
Macon, Chablis, Pouilly white, Baunes, and Lancel wines, as well as
eau de vie and liqueurs. In the ground-floor kitchen was a wide and
plentiful supply of cookware, including pots, kettles, and frying pans
made of red copper, and articles made of less valuable materials like
cast iron, tin, and iron.

Most of the dishes were kept in the dining room, also on the
ground floor, in a cupboard or in the marble-topped oak buffet. The
dining room was well lit, with two windows and two table lamps,
and heated with a stove.[6] A walnut dining table, two walnut chairs,
straw mats on the floor, several vases and platters, an oeil de boeuf,
and a liquor cabinet completed the furnishings. It appears that the
stove was shared with the adjoining, windowless room that served as
an office or study, presumably for Monsieur Gruyer.

The office was sparsely and cheaply furnished, with not much
more than a wooden board forming a desk and painted wooden
shelves to hold files and books, along with a few chairs, a lamp, and
two prints on the wall. It is noteworthy that the Gruyer household
had a sizable book collection, more than most small shopkeepers and
entrepreneurs, though not as extensive, varied, or valuable as the
libraries of merchants, professionals, and high-level civil servants.[7]
The Gruyer taste in books ran to the concrete—several histories of the
French Revolution and travel books. But the collection also included
unspecified works by Rousseau and a copy of *Robinson Crusoe,* among
other novels.

5. Archives Nationales, Minutier Central, LVI 657 (11 September 1829).

6. Roger-Henri Guerrand notes that over the course of the nineteenth century
the dining room became less intimate and more formal, and the family tended to
leave the dining room after meals and sit in the more comfortable salon. "Espaces
privés," in Perrot et al., *De la Révolution,* 333.

7. Daumard finds that professionals and bureaucrats were more likely than
other bourgeois to have a library. *Bourgeoisie,* 138. My own sample only weakly
supports this, especially for the households of the Restoration period. See the
discussion later in this chapter.

The Gruyer family, like other petty shopkeepers, slept on the upper floors, above the kitchen, dining room, and office. The largest and most richly furnished room upstairs was a combined salon and bedroom where Monsieur and Madame Gruyer slept. This was a common arrangement among families of the lesser bourgeoisie where a man's work required residence near a shop, or required that a significant room in the apartment be given over to professional purposes. For example, doctors, lawyers, architects, artists, and notaries often had their offices in the home, and unless they were extremely successful the home office was at the expense of a separate salon. Especially for the Gruyers, with three young children, and a business downstairs, space was at a premium. Nonetheless, in furnishing the salon/bedroom, the Gruyers displayed proper style, if not good taste, in the materials used. The bed, occupying an alcove that could be curtained off from the rest of the room, was richly and substantially draped in striped cotton fabric decorated with green silk fringe and fancy braid. In a tasteful manner (according to prescriptive literature of the time), the bed curtains matched the window curtains; and both the bed, with its high backs and curved sides in the form of a boat, and the marble-topped bedside table were made of rich, dark mahogany. Madame and Monsieur Gruyer slept on a straw mattress and two woolen mattresses, with an additional layer of feather bedding. The bedside table probably contained a chamber pot, and all furnishings having to do with sleep and other intimate functions could be discreetly hidden in the alcove during the day.[8]

Outside the sleeping alcove, the dresser, the round-top desk, two curved-back chairs, two armchairs, and six armless chairs were also made of mahogany, covered in the same striped cotton as the drapery, and decorated with green fancy braid. Mahogany furniture with matching upholstery was exactly what Madame Pariset advised in her handbook of housekeeping.[9] The other furnishings of this part of the room also conformed to various guidelines of appropriate bourgeois

8. "If a bedroom must serve simultaneously as a salon, it is tacky, even indecent, to want to *decorate* it with furnishings that are only for use at night or for dressing; thus one hides during the day the bedside table, the night lamp, the pillows, the washstand, et cetera, and one adds a little more elegance to the whole ensemble." Garnier-Audiger, *Manuel du tapissier, décorateur, et marchand de meubles* (Paris: Roret, 1830), 104.

9. Mme Pariset, *Nouveau Manuel complet de la maîtresse de maison* (Paris: Roret, 1852), 15, 18.

style. Around the fireplace were andirons, a shovel, and tongs made
of copper. Above the hearth was a large wall mirror framed in gilded
wood, and another similar but smaller mirror (probably facing the
first) was located above the dresser. The presence and locations of
these mirrors were standard for bourgeois households; perhaps they
were intended to be reminiscent of eighteenth-century aristocratic
practice or to give the illusion of light and space in apartments that
were more often than not dark and cramped.[10] In front of the mirror
and on the mantelpiece were other items common to bourgeois
homes—a pair of gilded and bronzed copper candlesticks and a mantel clock made of mahogany and gilded copper, all kept under glass
bells and resting on stands of darkened wood.[11] The Gruyers also
kept a tea set in this room, as well as two breakfast bowls made of
porcelain. Serving tea to guests in the salon was a socially acceptable
thing to do; but does the presence of bowls indicate that Monsieur
and Madame took their breakfast in bed or in the salon? This seems
likely, especially since they, like most members of their class, had a
household servant to perform such chores as preparing and serving
breakfast.

The Gruyers' modest collection of art and art objects set them
apart from other families of similar occupation and living quarters,
who showed less interest in art; it also aligned them with the bourgeoisie as a whole, searching for status through the appreciation and
consumption of art. They hung several prints representing battle
scenes on the wall, interspersed with four oil paintings of landscapes.
Statuettes made of copper or porcelain also decorated the room.

The children slept in a room furnished simply with a walnut,
web-bottomed bed and three woolen mattresses, a bedside rug, two
small rush-seated chairs, and a closet. Yet another room, on the top
floor, contained chairs, a dresser, a roll-top desk, engravings, and a
washing pitcher and bowl. It is difficult to determine just what
purpose this room served for the family. It was undoubtedly small,
without windows, and contained the only washing implements inventoried, apart from a child's bathtub located in the servant's room.
The Gruyers seem not to have been overly concerned with hygiene
and cleanliness. Other families had dressing or wash rooms near the
bedroom, containing a washstand, a bidet, and/or a chair with a hole

10. See Garnier-Audiger, *Manuel du tapissier,* 101.
11. See ibid., 102.

in the seat and a removable chamber pot below (*chaise percée*); this room in the Gruyers' home was rather inconvenient for regular washing.

The servant had a separate room on the top floor. She slept on a wooden bed furnished with two straw mattresses and two woolen mattresses, as well as bed sheets, a blanket, and a bolster. These accommodations were typical for servants during the entire nineteenth century, though many of the Gruyers' contemporaries lodged their servants in closets near the kitchen or elsewhere within the family's living quarters.[12]

The total value of inventoried possessions was 4,407 francs. The Gruyers were by no means rich compared to others in the sample; but they managed to adhere to certain conventions of bourgeois status (the servant, the mantel clock and candlesticks, matching mahogany furniture and drapery, some art, and good wine), even without a salon and other amenities that their wealthier peers enjoyed. The sample includes a wide range of wealth, measured by the value of household possessions; the Second Empire households tended to be richer, at least in the matter of home furnishings. The following list shows the distribution of households according to the values of possessions inventoried.

	Restoration	*Second Empire*
3,000–6,000 francs	6	6
6,000–9,000 francs	8	3
Over 9,000 francs	2	7

Given the expansion and rising fortunes of the bourgeoisie during the nineteenth century, this distribution is not surprising.[13]

The range between the lowest and the highest value of household goods is great, especially for the Second Empire. In the latter sample, the lowest value was 3,602 francs for a household headed by Monsieur Perrin, a manufacturer of firearms, and the highest was 28,540 francs for the medical doctor Sudry's possessions. The range for the Restoration sample was from 3,168 francs, the value of property

12. Sorlin, *Société*, 155. Anne Martin-Fugier, *La Place des bonnes* (Paris: Grasset, 1979).

13. Sorlin, *Société*, 129; Daumard, *Bourgeoisie*, 217.

Table 1.   Distribution of Households according to
Occupation of Head of Household and Value of Household Possessions

	Restoration				Second Empire			
	3,000–6,000 fr.	6,000–9,000 fr.	Over 9,000 fr.	Total Households	3,000–6,000 fr.	6,000–9,000 fr.	Over 9,000 fr.	Total Households
Trade (wholesale and retail)	2	5	1	8	1	0	0	1
Industry	1	0	0	1	2	0	1	3
Property or investments	1	1	0	2	0	1	2	3
White collar (civil service and private)	1	2	0	3	3	0	1	4
Liberal professions	1	0	1	2	0	2	3	5
Total number of households				16				16

owner Lalo's goods, to 17,544 francs for the possessions of the lawyer Delondre.

The sample indicates few discernible correlations between occupation and value of household possessions. Table 1 showing occupations and values suggests that whereas retail and wholesale trade were common and successful occupations under the Restoration, under the Second Empire the liberal professions were becoming more rewarding and prestigious.[14] But since the sample includes the value only of household possessions, not the family's total wealth, it is appropriate, in considering the role of consumption for the bourgeoisie during industrialization, to concentrate on the details of the inventories of goods rather than their total value. Just what were bourgeois homes like in these two periods, and how were they changing or staying the same?

SPATIAL DIVISIONS IN THE BOURGEOIS HOME

A distinguishing feature of the thirty-two households in this sample is the multiplicity of rooms in each. This suggests a degree of comfort

14. Daumard asserts that by the end of the July Monarchy newcomers to Paris had less chance of ascending to bourgeois status through shopkeeping or self-employed artisan manufacture. Adeline Daumard, *Les Bourgeois et la bourgeoisie en France depuis 1815* (Paris: Aubier, 1987), 135.

and a desire to separate physically the functions of private life from those of public life. The number of rooms, and therefore the divisions of function, were more numerous in 1870 than in 1830, suggesting that bourgeois families placed more importance on this spatial separation of functions and/or had more means to implement it. In addition, there were more children in the later sample than in the first—thirty-five compared to twenty-four—so it is not surprising that those families had more second bedrooms than the families in the Restoration sample. It is true, too, that there were more, and larger, apartments available in the later period; construction had been almost constantly booming in Paris during the nineteenth century, and many more apartment buildings intended for bourgeois occupants appeared in the center of the city during the urban renovation of Prefect Baron Haussmann.[15]

Almost all of the households had separate kitchens for cooking, dining rooms for eating, and bedrooms for sleeping.[16] Twenty-six households had at least one room set aside for entertaining—the salon. Five families in the earlier sample (including the Gruyers), and one family in the Second Empire sample, mingled sleeping and entertaining in one room—the combined bedroom and salon. The number and use of other rooms was more various in samples from both periods, with some noteworthy differences across time. For example, there were many more anterooms in the later sample than in the earlier one—twelve as opposed to five. An anteroom was the entry way into the home, a transition space between exterior and interior. It was furnished with, at a minimum, coat hooks, benches or chests, umbrella stands, and lanterns for outdoor use. Family members and guests took off and put on their outdoor gear in the anteroom; it was also a space where visitors might wait while masters and mistresses instructed their servants whether or not to admit them into the home.[17] The anteroom could also substitute for another room during populous stages in the family life-cycle (and thus the smaller size of Restoration families might partially explain the relative

15. Sorlin, *Société*, 138–39; Jeanne Gaillard, *Paris, la ville, 1852–1870* (Paris: H. Champion, 1977), 73.

16. Two inventories from the Restoration period omitted any mention of a kitchen. I cannot explain the omission, since both households were wealthy, and one listed a bedroom for the cook. Possibly the kitchens belonged to owners of the apartment and therefore were not inventoried. Notably, these were the only two households out of the entire sample with stables, a horse, and a carriage.

17. Guerrand, "Espaces," 332; Gaillard, *Paris,* 73.

absence of anterooms in 1828–30). Balzac describes such a substitution in *La Cousine Bette*. Madame Hulot, forced to move into a
smaller apartment because her husband the baron is squandering the
family fortune on his numerous mistresses, is put under additional
strain when her newly married daughter Hortense leaves her unfaith-
ful husband to return to the home of her parents. The two women
transform the dining room into a bedroom for Hortense and her
baby, while the anteroom becomes the new dining room.[18]

The increasing presence of anterooms in bourgeois homes expanded the amount of public space, as well as separating entry and
departure rituals from other functions. The amount of private space
also expanded; there were eleven second bedrooms in the sample
from the Second Empire, compared to seven in the sample from the
Restoration. This increase supports the interpretation of a growing
concern within the bourgeoisie for the separation of space and for
privacy.[19] However, the increase was also undoubtedly due to the
larger number of children in the later sample. Even so, cradles and
small children's beds were often located in the bedrooms of parents
and nurse maids, and it appears that boys and girls often slept two and
three to a "children's" bedroom.[20] It is often difficult to know exactly
where children slept and in how many beds, because in cases where
the parents stipulated that beds and other furnishings belonged to the
children, these items were not inventoried.

Another possible explanation for the greater number of second
bedrooms in the later period compared with the Restoration was the
declining number of rooms used as offices. In the sample from the
Restoration eleven households had an office, whereas only eight
households had business-related rooms in the Second Empire sample.
Historians have noted that the separation of the workplace and the
home was a significant trend in the nineteenth century, affecting both
working-class and bourgeois families but in different ways.[21] For the

18. Honoré de Balzac, *La Cousine Bette,* in vol. 17 of *Oeuvres complètes de M.
de Balzac* (Paris: Les Bibliophiles de l'originale, 1968).

19. Alain Corbin, *The Foul and the Fragrant* (Cambridge: Harvard University
Press, 1986), 164.

20. Michelle Perrot perhaps exaggerates the "sanctity" of the conjugal bedroom when she asserts that children were never admitted inside. At least when
children were young, it was not unusual for them to sleep in the same room with
their parents. Michelle Perrot, "Fonctions de la famille," in Perrot et al., *De la
Révolution,* 115–16.

21. Bonnie G. Smith, *Ladies of the Leisure Class: The Bourgeoises of Northern
France in the Nineteenth Century* (Princeton: Princeton University Press, 1981);

bourgeois families represented in this sample, the removal of business-related rooms from the home allowed for more strictly domestic space and perhaps greater comfort and convenience in the home. As more men entered manufacturing as owners and/or managers, and the number of civil servants grew, the likelihood of men needing rooms in the home for work purposes diminished. With the expansion of industry in terms of size and the amount of capital needed to launch an operation, fewer owners and managers lived alongside their businesses in the Second Empire compared to the Restoration.[22]

Yet another indication of increasing domestic space and privacy in the bourgeois household is the number of separate rooms for servants. In the earlier sample ten households had separate rooms for servants, compared to fourteen in the Second Empire sample. The increase is not great, but it does bear out the observation that bourgeois families were more inclined to separate themselves from servants as the century progressed, and that bourgeois apartments were tending away from the warren of closets, alcoves, and tiny rooms of earlier periods toward larger rooms with more specific purposes.[23]

Finally, a remarkable development is a decline in the number of rooms designated for washing or dressing or other bodily functions and in the number of implements for these purposes. In the earlier sample the inventories noted nine bidets, four toilet chairs (*chaises percées*), one hip-bath, three children's bathtubs, three foot baths, thirteen pitchers and bowls, and four washstands (which included pitchers and bowls). In the later sample there were only two bidets, two toilet chairs, one child's bath, one foot bath, seven pitchers and bowls, and three washstands. It is possible that assessors in the later period ignored these implements as having relatively little value. Possibly, too, there were more latrines belonging to the owners of apartment buildings in the Second Empire, diminishing the need for toilet chairs and chamber pots. But the numbers do suggest a lack of concern for personal hygiene.[24]

Erna Olafson Hellerstein, Leslie Parker Hume, and Karen M. Offen, eds., *Victorian Women: A Documentary Account of Women's Lives in Nineteenth-Century England, France, and the United States* (Stanford: Stanford University Press, 1981); Louise A. Tilly and Joan W. Scott, *Women, Work, and Family* (New York: Holt, Rinehart and Winston, 1978).

22. Sorlin, *Société*, 138.
23. Gaillard, *Paris*, 73; Guerrand, "Espaces," 332.
24. Sorlin, *Société*, 154–55; Corbin, *Foul*, 172–75.

The installation of running water in private apartments moved slowly during the nineteenth century; it was not commonly available to upper-floor dwellings on the right bank in Paris until 1865, and on the left bank not until 1875.[25] The exhibition reports had little to say on devices for, much less improvements in, human waste removal and personal hygiene; though Louis Wolowski explained that soap manufacturing was advancing in terms of both price and quality of products due to scientific research in chemistry.[26] Flush toilets were not recognized as important for hygiene or civilization by health experts in France until the 1880s.[27] All the more extraordinary, then, was Renée Saccard's fabulous, seductive bathroom so vividly described in Zola's *La Curée,* a novel about bourgeois excess and greed under the Second Empire. Renée, the wife of an ambitious upstart who makes a killing on land deals during Haussmannization, bathes every morning in a shell-shaped sunken bathtub of pink marble, with silver spigots in the shape of swan's necks.[28] This is a far cry from the washstands and bidets of less exalted Parisian bourgeois.

PUBLIC AND PRIVATE LIFE

Between 1830 and 1870 the number of households with salons increased from eleven to fifteen. This goes along with the general trend toward more rooms and the allocation of functions to specifically designated spaces. More striking, however, is the tendency of families to spend more on furnishing the salon, as compared to the bedroom, over the forty years between the two samples. I calculated for each household the percentages of the total value of possessions that different rooms represented. The following list shows the average figures for the two periods.

	Restoration	*Second Empire*
Salon	11%	20%
Bedroom	13%	12%
Dining room	5%	6%

25. Guerrand, "Espaces," 337.
26. Commission française sur l'Industrie des Nations, *Exposition universelle de 1851: Travaux de la Commission française sur l'Industrie des Nations* (Paris: Imprimerie impériale, 1855) 7:1–46.
27. Guerrand, "Espaces," 338.
28. Émile Zola, *La Curée* (1871; Paris: Fasquelle, 1984), 198–99.

This development suggests that over the nineteenth century bourgeois families placed increasing value on the living room as a sign of bourgeois status.[29] The living room was the most public room of the household, in the sense that this was where guests were entertained. The salon invariably contained a fireplace, often the only fireplace outside of the bedroom; it was therefore one of the most comfortable rooms in the home, especially in winter. The fireplace was also the site of some of the most expensive household furnishings—a mantel clock, bronze or gilded candelabras or sconces, porcelain vases, a mirror over the mantelpiece, and bronze or gilded metalwork outlining the hearth. If the family were rich or cultivated enough to own valuable prints, paintings, or art objects, these would be placed in the living room, where they could impress any outsider who visited the home with the wealth and taste of the household. As the century progressed, visiting among the bourgeoisie increased and definite rituals evolved for specific occasions.[30] Though bourgeois families often entertained guests and relatives for dinner in the dining room, hospitality for those outside the immediate family circle far more often occurred in the living room, with New Year's visits, courtships, balls or singing parties, and most common and frequent of all, weekly at-homes.

The development of the salon as a place of public display and entertainment coincided with the delegation of the responsibility for home decoration and hospitality to women. As noted in chapter 2, the decoration and maintenance of the salon for the reception of family, friends, and professional acquaintances were serious responsibilities that fell upon the mistress of the household.[31] At a dinner party at her cousin's house in Lille in 1865, the style and furnishings of Marie Wallaert's salon made a much deeper impression on Caroline Brame than did the company.[32] Similarly, Geneviève Breton, a

29. An 1830 guide to home decorating asserted that the salon was the real showpiece of the home; but families did not necessarily practice this principle until several decades later. Garnier-Audiger, *Manuel du Tapissier*, 101. See also Sorlin, *Société*, 154; Guerrand, "Espaces," 334.

30. Whitney Walton, "Feminine Hospitality in the Bourgeois Home of Nineteenth-Century Paris," *Proceedings of the Western Society for French History* 14 (1987): 197–203. Also see the discussion later in this chapter.

31. See Anne Martin-Fugier, "Les Rites de la vie privée bourgeoise," in Perrot et al., *De la Révolution*, 260; Anne Martin-Fugier, *La Bourgeoise* (Paris: Grasset, 1983), 187–93.

32. See chapter 2 for the full diary entry. *Le Journal intime de Caroline B.*, ed. Michelle Perrot and Georges Ribeill (Paris: Montalba, 1985), 46.

contemporary of Brame's and a resident of Paris, noted in her diary the sincere gratification her mother received in entertaining guests because she could give others pleasure through her lovely home and good table. "At heart [my mother] is very hospitable; she likes to have people like her house, . . . the abundance of her table, the order and keeping of her garden and of her interior."[33]

Much in contrast to these women who successfully maintained a tasteful and agreeable home atmosphere is the fictional Angélique de Granville, the pious, hypocritical, and unloving wife of a counselor at the Palais de Justice, in Balzac's *Une Double Famille.* Young Roger de Granville, busy with his new and prestigious appointment, leaves to Angélique the decoration of the newlyweds' Paris apartment—a task that is "usually a source of pleasure and tender recollection to young wives." But the result in this case is utter disaster.

> Whether it was that Madame de Granville had given her custom to tradesmen without any taste, or that her own nature was inscribed on the quantity of things ordered by her, certain it is that the young husband was astonished at the dreariness and cold solemnity that reigned in the new home. He saw nothing graceful; all was discord; no pleasure was granted to the eye. . . .
>
> If the slightest harmony had reigned, if the articles of furniture had taken, in modern mahogany, the twisted forms brought into fashion by the corrupted taste of Boucher, Angélique's house would merely have offered the odd contrast of young people living in the nineteenth century as if they belonged to the eighteenth; but no,—a mass of heterogeneous things produced the most ridiculous anachronisms. . . . Greek helmets, Roman broadswords, shields which, due to military enthusiasm, now decorated the most pacific articles of furniture, were little in accordance with the delicate and drawn-out arabesques, the delight of Madame de Pompadour.[34]

According to Balzac, the woman's poor taste and discordant furnishings are a reflection of her narrow-minded and prudish character. In the story this lack of taste is the first sign to Granville that his marriage was a mistake; and certainly he can never invite guests into the couple's inhospitable apartment. Angélique has failed to make her

33. Geneviève Breton, *Journal, 1867–1871* (Paris: Ramsay, 1985), 36.
34. Honoré de Balzac, *Une Double Famille,* in vol. 1 of *Oeuvres complètes,* 291–92.

terrain—the public rooms of the home and especially the salon—an environment that does credit to her taste and, by extension, to the family. In Balzac's fiction, and in the feminine press, the salon, along with the bedroom and the boudoir, was a "feminine" room in the sense that women furnished it and presided over it.[35]

The increasing value of salons compared to bedrooms suggests the priority of public display over private comfort in bourgeois households; but comfort was certainly not ignored. Indeed, both public and private aspects of home life were becoming more comfortable, due in part to technological changes but also to the ideal of domesticity.

THE LAYING ON OF FABRIC

One sign of increasing comfort that was also a symbol of status and wealth was the greater amount of and refinement in textiles used for drapery, furniture covering, and rugs. With the rapid expansion of textile production, the upholsterer gained prominence and a wider range of functions. Not only was he responsible for covering and upholstering furniture and for hanging drapery, but he also became an interior decorator, suggesting to clients styles and materials for furnishings, and procuring them.[36] As with cooking, interior decoration was a skill that all women were expected to cultivate; but professional decorators, like great chefs, were usually men. In Balzac's novel *César Birotteau,* Braschon, the wealthy upholsterer and furniture dealer, helps the architect Grindot remodel the home of César Birotteau, the successful perfume maker.[37] In 1830 Roret, the publisher of numerous how-to books, released the *Manuel du tapissier, décorateur et marchand de meubles.* One of the most important skills of the upholsterer that the manual emphasized was a thorough knowledge of fabrics—their composition, their properties, and how to

35. "Masculine" rooms included the study (the study is the only room that Granville furnishes himself), the billiard room, and the husband's bedroom if the couple did not share a room. The dining room was more gender neutral than any other room. For the gendering of rooms and pieces of furniture, see Leora Auslander, "Women Subjects and Feminine Objects: Women as Consumers and as Images in Late Nineteenth-Century Paris" (Paper presented at the Seventh Berkshire Conference on the History of Women, Wellesley, Mass., 19–21 June 1987).

36. Mme Pariset, *Manuel de la maîtresse de maison,* 3d ed. (Paris: Audot, 1825), 52.

37. Honoré de Balzac, *César Birotteau,* vol. 10 of *Oeuvres complètes.*

drape and combine them in order to furnish a room tastefully.[38] Whether consumers employed these services of the upholsterer or not, furniture upholstery, its covering, and drapery were essential aspects of furnishing.

In the Restoration inventories, all recorded dining-room chairs had seats of rush; rush chairs were also found in other rooms of the household. But chairs and sofas in the salon were generally more richly upholstered and covered. A common fabric for covering furniture during this period was Utrecht velvet, made of linen and goat hair. This fabric was sufficiently sturdy, inexpensive, and good-looking to make it very popular among bourgeois householders.[39] The other common material for covering furniture was wool; however, desk chairs might be covered with leather or more often baize. Horsehair furniture appeared in several offices, salons, and bedrooms. The wealthiest family in the early sample had a matching set of salon furniture covered in red silk.

In most rooms with windows, especially important rooms like the salon, bedroom, and dining room, a double set of curtains shielded the home from light, cold, and the outside. Almost all windows had at least muslin curtains hung across them, often accompanied by outer curtains made of heavier and richer material. In the early sample such outer curtains were almost all made of calico or other types of cotton fabric. As for the floor covering, occasionally families had small Aubusson rugs—an expensive and prestigious item—but usually another, less costly type of rug lay underfoot in selected spots, like beside the bed or in front of the fireplace.

Fabrics were more varied and more abundant in the later period, and chair coverings also changed dramatically (figs. 10 and 11). By this time, cane had replaced rush as the most common seat on armless chairs; most dining rooms had cane chairs, except in the wealthiest households, where people sat more comfortably or elegantly on moleskin, leather, or Moroccan leather. Rush chairs were largely relegated to kitchens, servants' quarters, and children's rooms. Velvet, probably of wool, was a common fabric for covering furniture,

38. Garnier-Audiger, *Manuel du tapissier*, 3. Cf. William M. Reddy, "The Structure of a Cultural Crisis: Thinking about Cloth in France before and after the Revolution," in *The Social Life of Things: Commodities in Cultural Perspective,* ed. Arjun Appadurai (New York: Cambridge University Press, 1986), 261–84.

39. Garnier-Audiger, *Manuel du tapissier,* 18; Pariset, *Manuel,* 33.

usually in rich colors like red, garnet, magenta, or occasionally green or blue. Damask appeared in several households, as well as rep, a combination of silk and wool, wool and cotton, or just cotton. Wealthy families had their furniture covered in silk, silk damask, or tapestry. Occasionally furniture was extravagantly padded, and a few households had fashionable poufs—ottomans that consisted only of padding with no wooden framework at all. Curtain fabrics were also more varied under the Second Empire. Not one household inventory indicated calico curtains; instead common fabrics were rep, wool and silk damask, and silk, usually matching the covering on the furniture.

In the later sample more households had rugs and room-sized carpets. The taste for and accessibility of rugs had increased steadily over the nineteenth century, and the author of the upholsterer's guidebook called carpeted floors "the height of sumptuousness of even the most magnificent furnishings."[40] Floor rugs were commonly made of linen and wool. The most expensive and durable rugs, owned by wealthier families, were from Turkey; some householders bought imitation oriental rugs. Popular, too, were the skins of wild animals—fox, lynx, panther—on top of beds or an added layer on a carpet. Table rugs—covering game tables, dining tables when not in use, and pedestaled tables—were ubiquitous. Constance Aubert's fondness for padding and soft coverings over all hard surfaces[41] was shared by many Second Empire households.

Rich and abundant fabrics represented wealth and status; to have a home furnished with silk and wool cloth, rugs, and thick curtains combined both the luxury of the aristocratic past and the contemporary sense of domestic comfort. Industrialization, by increasing textile output and lowering the cost of fabric, and by creating new dyes and new textile blends, allowed more bourgeois families to achieve a higher degree of status and comfort through the purchase of more and varied fabrics for household furnishing.

BEDS, STYLE, AND COMFORT

Another area where comfort increased between the two sample periods was in bedding. Beds and bedding underwent significant changes during the middle decades of the nineteenth century in

40. Garnier-Audiger, *Manuel du tapissier,* 142.
41. See chapter 2.

France, due largely to the increasing availability and decreasing price of iron. Yet for the outer appearance of the conjugal bed consumers upheld the traditional use of wood as the basic raw material, favoring ever larger and more massive examples of the woodworker's art.

A conjugal bed, usually bought at the time of the couple's marriage, was intended to last a lifetime.[42] From the Restoration through the Second Empire, beds were solid and stable pieces of furniture, constructed of good wood, often mahogany; and they were frequently hung with draperies for warmth and discretion. The conjugal bed was symbolic of the permanent union that bourgeois marriages ideally represented. Also, it reflected social status as much as did the more public furnishings in the salon. Throughout the nineteenth century, wooden beds usually had tall headboards and footboards and fairly wide sides, all resting upon thick feet (fig. 12); moving the bed was difficult. Another popular form for the wooden bed during the Restoration period was called a boat bed (*lit à bateau*), where the sides of the bed curved upward to meet the headboard and footboard, further enclosing the sleeping couple in a separate space, a ship sailing in a domestic sea. It is noteworthy that in the two samples, all beds shared by a married couple and, in the two cases where there were separate bedrooms, the bed located in Madame's room, were made of wood.[43] The total number of wooden beds in the Restoration sample is greater than that in the Second Empire sample—thirty-eight compared to twenty-eight—but this is because the count includes beds belonging to children and servants, which in the later sample were usually made of iron.

Iron beds were still a fairly recent innovation at the time of the Crystal Palace exhibition, especially for domestic use (fig. 13). During the July Monarchy (1830–48) doctors and public health experts recommended iron beds for use in hospitals and other public institutions, for hygienic reasons.[44] Iron beds were not subject to termites or

42. Daumard, *Bourgeoisie*, 136.

43. The inventories that designated separate bedrooms for husband and wife were those of the two most richly furnished households in the Restoration sample. In both cases the wife's bed and bedroom were more lavish and expensive than the husband's.

44. Corbin, *Foul*, 99. Iron beds were also being adopted as furnishing by the poor. Musée Industriel, *Description complète de l'Exposition générale des produits de l'industrie française faite en 1834* (Paris: Société polytechnique et du recueil industriel, 1836) 2:351.

woodworm, they were easier to clean and disinfect, they were durable and mobile, and they were becoming cheaper. They were clearly a consequence of the Industrial Revolution; in the sample of inventories from the Restoration period there was only one iron bed listed, whereas the Second Empire sample included thirty-two iron beds, all of them in servants' and children's bedrooms.[45]

This distribution makes sense for status-conscious but practical bourgeois consumers. For the symbol of wealth and domesticity—the conjugal bed—householders continued to choose beautiful, big, permanent, more or less handmade wooden beds. But for servants and children—transitory, socially inferior beings whose rooms were rarely on public display—the new iron beds were practical and economical. Though exhibitors at the Crystal Palace tried to appeal to consumers' sense of taste and style by displaying wrought-iron beds that imitated wooden beds in size, solidity, and craftsmanship (fig. 14), this attempt obviously failed. The properties of iron that bourgeois householders appreciated as a material for beds were low cost, durability, mobility, and perhaps cleanliness. But iron was no substitute for the lustrous finish, the solidity, tradition, and status of a mahogany or rosewood bed.[46]

While the outward appearance and material of conjugal beds did not change significantly from the Restoration to the Second Empire, the internal composition of bedding did develop dramatically. Under the Restoration, and for centuries earlier, a common method of constructing wooden beds was to nail rope or other strong material across the bottom to form a woven or webbed surface. Upon this webbing householders piled several mattresses, often putting a straw-filled mattress directly on the webbing, with one or two (sometimes even three) wool-filled mattresses on top. In the sample from the Restoration, thirty-eight beds had webbed bottoms, twenty-three of them used by servants. During this period a basic wooden frame of pine or another lesser wood, strung with webbing, constituted the cheapest sort of bed. Even married couples of the Restoration

45. One iron bed described as folding and located in the parents' bedroom was probably for a child.

46. In his analysis of exhibits from the 1851 and 1855 Exhibitions Charles Laboulaye mentioned that iron furniture was cheaper, but not as elegant as wood. Charles Laboulaye, *Essai sur l'art industriel* (Paris: Bureau du dictionnaire des arts et manufactures, 1856), 87.

bourgeoisie slept on webbed foundations and straw mattresses, but horsehair mattresses were also available and adopted for the conjugal bed, no doubt at higher cost.[47] Sixteen horsehair mattresses appeared in the inventories from the Restoration. In general, these represented an improvement over the straw mattress and webbed foundation in terms of comfort and hygiene.[48]

A significant change in bedding material, with implications for bed construction, was the development and adoption of the box-spring mattress between the Restoration and the Second Empire. Like the iron bed, the box-spring mattress was a product of modern industrialization, specifically of the increased output and falling price of iron. Whereas no households in the Restoration sample had a box-spring mattress, nineteen of them were inventoried in the selection from the Second Empire. A women's magazine stated in 1851, "Box spring mattresses are the most sought after; they are currently, so to speak, in style."[49] Box-spring mattresses made webbed bottoms unnecessary, and only three beds in the Second Empire sample were noted as having them. Instead, box-spring mattresses fitted into frames on wooden and especially iron beds, and they required fewer mattresses piled one atop the other for comfort. The modern practice of placing one stuffed mattress above a box-spring one was common in Second Empire Paris. The same number of straw mattresses appeared in both samples; since there were more beds in the later sample and since nineteen of them had box springs, this figure represents a decline in use, especially for the beds of married couples. As with iron beds, straw mattresses were more often used for the beds of children and servants than for those where the master and mistress of the household slept.

Bourgeois consumers took advantage of industrial developments to increase their sleeping comfort, as long as comfort did not interfere with the desire for status and display in the appearance of the conjugal bed. Bourgeois householders generally enjoyed greater comfort; they

47. It is difficult to distinguish the cost of webbed bottoms and straw mattresses compared to box mattresses because the inventories almost always assess all the bedding together, including several wool and feather mattresses, pillows, bolsters, and sometimes sheets, blankets, and bedspreads.

48. Garnier-Audiger, *Manuel du tapissier,* 49.

49. *Le Conseiller des dames,* March 1851, 155.

provided more family members with their own, individual beds; they economized on servants' accommodations; and still they upheld prestige and taste in the furnishing of the main bedroom.

HEATING AND LIGHTING

Industrial developments also affected heating and lighting devices. Yet here again, bourgeois concern for the appearance of wealth, status, and taste in the home moderated interest in cheap, efficient, modern methods of heating and lighting. For centuries, fires in fireplaces had heated residences and had provided an important source of light. Candles made of wax or tallow, and simple lamps— hardly more than wicks dipped in oil contained in a metal receptacle—were the only other home lighting devices. The nineteenth century in France witnessed great advances in the efficiency, cleanliness, and economy of utensils providing heat and light for the home; but the selectivity with which bourgeois households adopted these new devices reflects priorities of taste and status.

A significant continuity between the earlier sample and the later one is the fireplace as the main source of heat in the home. Indeed, there were more fireplaces in the households from the later period, despite technological advances in alternative forms of home heating. Fireplaces themselves were never inventoried, since they were part of the building's structure rather than a personal possession of the occupants; but the mention of andirons, shovels, tongs, brooms, bellows, and fenders in the inventory of a room signaled the presence of an operating fireplace. There were forty-two fireplaces in the Second Empire group of households, compared to thirty-two in the Restoration sample.

Several explanations for these figures are possible. The greater number of fireplaces may reflect a higher standard of living among Second Empire bourgeois; as noted earlier in this chapter, the apartments in that period had more rooms than those from the Restoration. Though the studies of Haussmannization under the Second Empire say little about the size and furnishings of the new apartment buildings constructed for bourgeois residents, it is likely that these new (and expensive) edifices contained more rooms per apartment and more fireplaces per room than the older, cramped apartments of the early nineteenth century.[50] But a cultural and social interpretation of the persistence of, even increase in, fireplaces in bourgeois households is persuasive, given the locations of fireplaces throughout the home, the additional possessions that fireplaces entailed, and the prosaic appearance of stoves.

In the sample from the Restoration the room that most frequently contained a fireplace was the bedroom—suggesting minimal social activity on the part of the family, or a priority of private comfort over public display. The second most frequently heated room in the early period was the salon; nine out of sixteen households had fireplaces in their salons. Other bedrooms, dining rooms, and other rooms were much less frequently heated by fireplaces or even stoves, though households often had bed warmers for localized warmth. In the later period a slight shift toward more heating of salons than of bedrooms corresponds with the trend toward more family sociability and the importance of the salon as a room to display family wealth and taste. In the later sample thirteen fireplaces were located in salons, and eleven in the bedrooms of the heads of the households. A concern for both private and public comfort, however, is evident in the fact that fireplaces were located in eight additional bedrooms and eight home offices. More children and guests enjoyed the benefits of fireplace heating in the rooms where they slept, and men who worked at home were warmer and could accommodate their clients and associates more agreeably and impressively.

Stoves as a source of room heating were remarkably underutilized in mid-nineteenth-century France. Indeed, the number of stoves in the sample declined from five in the Restoration to one in the Second Empire, and that one was located in a servant's room. Stoves were unquestionably a much more efficient and cheaper means of heating than fireplaces, since their principle of operation was to heat air through contact with the walls of the stove instead of merely radiating heat from a fire. As soon as air surrounding the stove was heated and rose, new, cooler air replacing the warm air would be heated, and so the air in the entire room would be evenly heated and reheated. Stoves eliminated the problem of poor draw and the consequent

50. Sorlin, *Société,* 138–39; Gaillard notes the appearance of central heating in new bourgeois apartment buildings of the Second Empire, though she suggests the devices may not have been totally effective. Gaillard, *Paris,* 72.

invasion of the room by smoke and gas.[51] Prescriptive literature condoned the use of tile or metal stoves, especially in the dining room; but out of the thirty-two households, only one (the Gruyers from the Restoration) heated the dining room with a stove. Rarely did dining rooms have fireplaces (two in the earlier group and one in the sample from the Second Empire).

Why did stoves for home heating—presumably a modern technological advancement in an industrial age—decline in use among bourgeois householders, while the number of fireplaces increased? Consider their respective appearances. Metal stoves could be squat metal boxes placed against a wall, in a bare fireplace, or at a distance from the wall, with a long metal pipe that carried smoke up through the ceiling and out of the room (fig. 15). Though aesthetically ungainly, they were more effective than traditional fireplaces.[52] Experts applauded the increasing efficiency and declining cost of stoves made in France; but they noted French stubbornness in favoring the least efficient, least modern, and most expensive method of home heating—the fireplace. "It is the desire to see the fire, to be able to arrange it, to poke it, that has for a long time opposed the improvement of heating implements."[53] Others hinted that Parisian householders did not really need the power and efficiency of stoves since the climate was milder than in northern European countries, where stoves were more widely used among persons of all classes.[54] What observers failed to mention was that stoves were simple and functional, whereas fireplaces embodied a variety of status and domestic connotations.

The fireplace was the site of very expensive articles to decorate the mantel and the hearth.[55] It was impossible to place a clock or a candelabra on top of a stove; this is probably an important reason why bourgeois consumers declined to use stoves, especially in living rooms and bedrooms. Moreover, the wealthiest householders had

51. Commission française, *Exposition* 6:14.

52. Ibid., 7.

53. Ibid.

54. Ibid., 15; Garnier-Audiger, *Manuel du tapissier,* 149.

55. The exhibition report notes that the English in particular liked to decorate their fireplaces with permanent fixtures made of marble, cast iron, or finer metal. By contrast, the French placed greater importance upon the movable objects of decoration—mantel clocks, candelabras, and so forth. Commission française, *Exposition* 6:53.

their fireplaces built of carved stone, marble, and especially cast iron work (fig. 16). The style and materials of these fireplace accoutrements were obvious signs of a family's wealth and taste.

In addition to being signs of prestige, fireplaces were the central focus of domestic unity and closeness. Families and friends congregated naturally around an open fire, but not around a stove.[56] The open fire, with its deep symbolism of the home and family, did not give way to the practical and less expensive stove in bourgeois homes. In this case, consumer preference for status and perhaps a domestic ideal prevailed over efficiency, thrift, and innovation. The consumption of lighting devices reflected similar priorities, but to a lesser extent.

While candles themselves did not appear in the inventories, every household had its candleholders, candlesticks, candelabras, sconces, and lanterns. A lantern enclosed a candle on three sides and top and bottom to protect it from wind and rain outdoors. Wax candles were the finest and most expensive kind; they burned brightest and with the least smoke or smell. Tallow candles were cheaper; but their light was somewhat dull and, being made from animal fat, they gave off an unpleasant odor. An important discovery hailed in French reports of the exhibition was the successful manufacture of stearic candles. These were made from tallow by a chemical process that eliminated the disadvantages of tallow candles. Stearic candles were more expensive than tallow candles, but not as costly as wax.[57] They were therefore capable of enhancing domestic comfort without excessive expense; and they still allowed householders to display their wealth and taste in the candlesticks, candelabras, and sconces they bought.

The following list shows the variety of lighting devices in bourgeois households in the two samples.

56. Wolfgang Schivelbusch suggests that peasants in the nineteenth century continued to use "archaic" open-flame lighting devices as a replacement for the open fire of the hearth. While peasants in the German countryside may have adopted stoves for heat and retained open-flame lamps or candles for light, Parisian bourgeois retained the open flame in both fireplaces and lighting devices. Wolfgang Schivelbusch, *Disenchanted Night: The Industrialization of Light in the Nineteenth Century,* trans. Angela Davies (Berkeley: University of California Press, 1988), 162–66.

57. Commission française, *Exposition* 7:56–68.

	Restoration	Second Empire
Candleholders	10	3
Candlesticks	63	67
Candelabras	51	31
Sconces	11	6
Lanterns	1	9
Chandelier	0	1
Carcel lamps	3	0
Moderator lamps	0	9
Hanging lamps	0	12
Unspecified lamps (manually regulated)	39	30

The cheapest and simplest lighting devices were candleholders (*bougeoirs*)—small dishes, often made of tin or another inexpensive metal, with a place for the candle and probably some kind of handle. These were often located in kitchens and were undoubtedly more functional than decorative. It is not surprising that the number of candleholders fell from the earlier to the later sample, since householders in the Second Empire owned more lamps to serve their basic lighting needs and they still possessed expensive candelabras and especially candlesticks for display. Candlesticks and candelabras were essential decoration for the mantel above the hearth; they were often made of bronze, or bronzed or gilded copper, and formed part of a set with the mantel clock. Wooden or metal sconces, usually gilded, were also common around the fireplace, though the sample suggests they were less popular under the Second Empire than during the Restoration. The greater number of lamps in households from the later period may explain the decline in wall sconces, which were less mobile and perhaps more difficult to coordinate with the style of a room. Lanterns were found exclusively in anterooms; their increase corresponds with the greater number of anterooms in the later sample, as well as the overall increase in the value of furnishings over the two periods. Only one household, from the richer Second Empire sample, possessed the supreme status symbol—a chandelier. Prescriptive literature favored a crystal chandelier with candles in the salon as the height of good taste;[58] but judging from this sample, it was a luxury few indulged in. Lacking this blaze of reflected light suspended from the ceiling, most bourgeois householders topped their fireplace mantels, chiffoniers, side tables, and walls with a combination of candlesticks, candelabras, sconces, and lamps. Candlelight, like fire heat, was associated with traditional, prestige articles of furnishing. "In general," wrote the author of the upholsterer's manual, "lighting by candles alone is the richest and the most distinguished for a sumptuous apartment."[59] Candlesticks, like fireplaces, were most commonly located in living rooms and bedrooms.

However, no household in the sample, even under the Restoration, was lit entirely by candles. Lamp technology was rapidly advancing in the nineteenth century, and lamps were becoming common in bourgeois households. Technological and artistic improvements made the light of table and mantel lamps brighter and cleaner, made them easier to use, and made them beautiful and decorative.[60] Improvements included new devices to regulate the flow of oil or gas to the lighted wick, new methods of gilding and bronzing to make lamps look beautiful even when made of base metals, and creative designs and artistic globes that rendered lamps as ornate and distinctive as candlesticks and candelabras. In the first half of the nineteenth century the richest bourgeois owned Carcel lamps, named after their inventor, who patented his device of regulating the oil flow to the lamp's wick with clock mechanisms in 1800. These lamps gave off a clear, clean, steady light, but the mechanism made them expensive and difficult to repair.[61] Cheaper and of simpler design were moderator lamps, invented in 1836. A spring put pressure on the oil, forcing it up to the wick when needed. Moderator lamps also shed good light, though not for as long as Carcel lamps. They were increasingly popular among the bourgeoisie, not least because they were highly decorated and were as much art objects as the candlesticks and candelabras with which they coexisted (fig. 17).[62] Indeed, the exhi-

58. Garnier-Audiger, *Manuel du tapissier*, 101.

59. Ibid.

60. Schivelbusch emphasizes that technological improvements in "old" forms of lighting delayed their obsolescence in the face of the truly revolutionary gas lighting. Schivelbusch, *Disenchanted Night*, 49–50.

61. Commission française, *Exposition* 6:78.

62. Ibid., 101. The feminine press expressed great interest in the aesthetic qualities of lamps: "I saw at this manufacturer's, bronze lamps with ornaments in

bition report on lighting blamed the fine workmanship and artistic beauty of French-produced lamps for their high price.[63] In 1851 French lamp makers were not catering to a mass market for cheap, functional articles; rather, they were satisfying the bourgeois desire for tasteful and artistic home furnishings.

An exception to this tendency was the hanging lamp (*la suspension*). No households in the Restoration sample had such a light, but twelve did in the sample from the Second Empire. Hanging lamps were especially popular in the dining room; not only did they bring light to what were often very dark locations, but they allowed people to dine more comfortably, without lamps or candlesticks taking up space on the table.[64] Illustrations suggest that these hanging lamps were more functional than beautiful, though some were made of gilded or bronzed copper and were thus valuable. They were not easy to use or to clean, since both operations required lowering the lamp and reaching it, often over the expanse of a dining table. Though it is difficult to determine from the inventories, placing candles on the dining table does not appear to have been a common practice. Almost the only time candlesticks or candelabras were inventoried in a dining room was when the dining room had a fireplace.

Based on the inventories, it is easy to conclude that dining rooms were in general poorly lit until the adoption of the hanging lamp. Possibly the absence of candlesticks and lamps in dining rooms, especially in the Restoration sample, was due to the timing of meals in French bourgeois homes. A main family meal in the middle of the afternoon would not have required much artificial light. As dinner times were pushed back later and later in the afternoon and into the evening during the course of the century, the desire for artificial light undoubtedly coincided with the availability of hanging lamps, and these basic items were added to dining-room furnishings.[65] It is also

---

relief that were perfectly modeled and finished. Others made of porcelain with Chinese paintings were worthy of attracting the attention of those who love to find in a useful object the always desirable union of good taste and elegance." *La Gazette des femmes,* 22 October 1842.

63. Commission française, *Exposition* 6:57–58.

64. Garnier-Audiger, *Manuel du tapissier,* 100.

65. During the eighteenth century dinner was served at 4 P.M. at the latest; but in the nineteenth century dinner times moved to between 5 and 6 P.M., to accommodate businessmen's responsibilities. Martin-Fugier, "Rites," 202.

possible that hanging lamps were preferable to candlesticks for family meals, where children and adults sat around a table small enough for intimacy and convenience—most dining tables inventoried had lengtheners to expand for guests and additional family members on special occasions.

In matters of lighting, then, bourgeois householders adopted the modern means—lamps—but, with the exception of hanging lamps, enforced standards of beauty and workmanship in their construction. Nor did they wholly abandon the traditional form of light—candles, with their rich, prestigious accoutrements and connotations of wealth and tradition. Again, the bourgeoisie was interested in comfort but still preferred some of the old forms to a modern, economical ease.

THE RISE OF CONSPICUOUS CONSUMPTION?

A different and new kind of consumer priority is evident in the increasing proportion of value represented by female clothes and jewelry from the Restoration to the Second Empire. Chapter 2 argued that nineteenth-century men and women believed that consumption was primarily a feminine function, complementary to (or subversive of) the masculine preoccupation with production. The inventories bear this out, even suggesting that Veblen's notion of conspicuous consumption also operated, at least in some fashion, among Parisian bourgeois. The value of female compared to male clothing in the inventories is a case in point.

In both the Restoration and Second Empire samples the percentage of spending on female clothes was higher than for male clothes, but the gap increased dramatically between the sample years. The following list shows the average values of male and female clothing, expressed as percentages of the total values of all the households' possessions.[66]

	Restoration	*Second Empire*
Female clothes	9%	11%
Male clothes	7%	4%

66. Cf. Marguerite Perrot, *Le Mode de vie des familles bourgeoises, 1873–1953,* 2d ed. (Paris: Presses de la fondation nationale des sciences politiques, 1982), 86–88.

It should be noted that the inventories do not clearly indicate how much of the clothing was acquired before or after marriage. However, men customarily gave their fiancées gifts of expensive jewelry, clothes, and even furniture on the eve of marriage. A fine cashmere shawl, for example, was an appropriate part of *la corbeille,* the groom's gifts to his bride; and this item was frequently the most valuable among the women's clothing in the inventories. Whether the clothes were purchased before, at, or after marriage, clearly more money was spent on female clothes than on garments for men.

Several historians have analyzed clothing in the nineteenth century, noting that appropriate male dress among the bourgeoisie was limited to a dark suit, while female clothes remained elaborate and varied, requiring custom tailoring and therefore being more expensive.[67] Fashion and social practice dictated that women buy different outfits and accessories for specific occasions and in rapidly changing styles. Was this a matter of conspicuous consumption? Did the increasing wealth of bourgeois families result in a desire to display social status in lavishly dressed women?[68]

This is a forceful explanation for the findings tabulated above, especially when considered with the different developments in clothing manufacturing for male and female consumers. To be sure, standardization of male dress and mass production of male clothing may have lowered the cost of men's clothes; but this hardly seems adequate as the sole explanation for the widening gap between the values of husbands' and wives' garments. Given the distinct ideals of feminine and masculine functions in bourgeois society, the growing importance of female dress probably contributed to the larger amounts spent on women's clothes. Men made their mark in society through their occupations; they relied relatively little on outward appearance to proclaim their social position (except dandies—but they would have fallen outside my criteria for inclusion in the sample). By contrast, clothing and household furnishings were the only

67. Henriette Vanier, *La Mode et ses métiers: Frivolités et luttes des classes, 1830–1870* (Paris: Armand Colin, 1960): Philippe Perrot, *Les Dessus et les dessous de la bourgeoisie* (Paris: Arthème Fayard, 1981); Whitney Walton, " 'To Triumph before Feminine Taste': Bourgeois Women's Consumption and Hand Methods of Production in Mid-Nineteenth-Century Paris," *Business History Review* 60 (1986): 541–63.

68. See Thorstein Veblen, *The Theory of the Leisure Class* (New York: Macmillan Co., 1899), esp. 80–85.

means for women to express social position or ambition. Conspicuous consumption, then, may have been as much an active goal of women asserting class and personal identity, as a passive one foisted upon them by men who themselves could not afford to live a life of leisure and frivolous spending. However, conspicuous consumption in mid-nineteenth-century France, tempered by lesser incomes and an aristocratic model of tastefulness, was not exactly the same as the phenomenon Veblen observed in the late-nineteenth-century United States. The French case suggests that gender, as well as class and economics, was a factor in the rise in spending on female finery, and that an understated tastefulness, rather than brazen conspicuousness, was the goal of this particular type of consumption.

Increased spending on female compared to male clothing was accompanied by greater spending on female jewelry, compared to household silver and linen. The following list shows the average values of jewelry, silver items (table flatware, teapots and coffeepots, serving spoons, tureens, bowls, and coffee spoons), and linen, expressed as percentages of the total values of the householders' possessions.[69]

	Restoration	Second Empire
Jewelry	19%	24%
Silver	21%	10%
Linen	11%	2%

The mention of diamond earrings in almost all of the inventories suggests that this item, like the cashmere shawl, was a widespread symbol of bourgeois status. Like shawls, diamond earrings and other types of jewelry may have come into a woman's possession as part of her *corbeille*. Bourgeois women possessed many pieces of jewelry made of gold, and rings and earrings set with semiprecious gems like garnets, coral, pearls, and turquoise were also popular; but often the most expensive item of jewelry in a household was a pair of diamond earrings, and in wealthy families this might be accompanied by a diamond necklace.

69. In one wealthy family, headed by the notary Panhard, jewelry amounted to 44 percent of the value of all household furnishings and possessions. Archives Nationales, Minutier Central, XLI 1134 (4 July 1870).

Like the increase in spending on female as opposed to male clothes, the higher spending on jewelry indicates a predominance of female taste in consumption. To be sure, males also had jewelry, and men's watches were an expensive bourgeois male accoutrement. Nonetheless, the amount and value of women's earrings, necklaces, bracelets, rings, hair combs, lorgnettes, and watches invariably exceeded those of male jewelry. Like women's clothes and more obviously than other articles of household furnishing, jewelry reflected the status of female consumers. Clothes and jewelry mirrored a woman's personality and desires, as well as her reading of current fashion and the symbols of social status.

There are other factors to account for in the shift of spending from silver and linen to jewelry between the Restoration and the Second Empire. One major technological advance that may have contributed to this trend was electrolytic metal plating (see chapter 5). With this new technique silver-plated goods, cheaper than the solid silver variety, became more abundant. This may have rendered silver tableware less prestigious in the minds of bourgeois consumers.

The decline in the value of household linen as part of a household's possessions may be due in part to urbanization. Bourgeois families had moved away from the rural and previously common practice of stockpiling large amounts of bed sheets, pillow cases, tablecloths, napkins, towels, aprons, and so forth.[70] Urban women bought these items, rather than making them as their rural forebears did; and factory production made household linen readily available.

No doubt it was women who chose to spend more on clothes and jewelry and less on linen and silver; this spending pattern reflects a development in social symbols and cultural norms emphasizing the style, taste, and richness of a bourgeois woman's appearance over the amount and value of her household linen and silver.

LEISURE, SOCIAL RITUAL, AND CONSUMERISM

Two significant features of bourgeois life in the nineteenth century were the opportunity for leisure and the growth of distinctly bourgeois forms of sociability. The inventories do not illuminate exactly how householders spent their time and entertained themselves, but

70. Daumard, *Bourgeoisie,* 135.

certain possessions suggest likely forms of entertainment and social ritual practiced in bourgeois homes. As with other aspects of bourgeois life, the taste for particular articles relating to leisure and sociability was related to developments in several areas of manufacturing, and to social and economic changes.

The most common possession connoting leisure activity in the bourgeois home was the game table or game box, located in the salon, bedroom, or dining room. Of the sixteen households from the Restoration period, eleven had at least one game table, and two had a game box. The interest in card playing and perhaps other types of indoor games persisted under the Second Empire; in this sample there were thirteen game tables, as well as a game box in one of these households. Although the term *table de jeu* refers to both card tables and game tables, it is likely that families used them primarily for cards, and perhaps for dominos (if they used the game tables for leisure at all); backgammon boards and chess sets were inventoried separately, and very few households possessed them—one household from each period had a backgammon board and one household from the earlier sample had a chess set.

Second to the game table in popularity as an article of furnishing for home activity, if not entertainment, was the work table or sewing table. The work table was a small table, sometimes very beautiful and intricate, with drawers, compartments, and/or a cloth bag in which to keep implements and materials for needlework. Six households from the Restoration sample and ten from the Second Empire had at least one work table, and two households in the earlier sample had work boxes for the same purpose. The work tables were undoubtedly purchased partly for show; in general small tables increased in popularity during the nineteenth century as displaying the craftsmanship of the producer and the good taste and refinement of the consumer.[71]

But work tables also reflected the widespread expectation that females of the bourgeoisie should be skilled in needlework. The feminine press emphasized fancy work—more decorative than useful sewing—as an attractive accomplishment for a girl or woman, consistent with the ideal of the female as embellisher of the home.

71. Roche, *People of Paris*, 148–49. One furniture dealer named Tahan, referred to frequently in exhibition reports and in the feminine press, built his reputation and fortune entirely on the manufacture of small decorative tables and other lesser pieces of furniture.

Nonetheless, women also did useful sewing, mending worn and ripped garments and sometimes making clothes for the family.[72] Thus the work table was not strictly an object of leisure in the bourgeois household, but a reflection of the feminine ideal in all its ambiguity. For the ideal maintained that bourgeois women be leisured and concentrate on activities that enhanced the comfort and beauty of the home. In practice, however, sewing was one of several means whereby women could economize on household spending (another such strategy was to skimp on daily meals) while still maintaining appearances with certain expensive items of furnishing and clothing. The work table was a fashionable piece of furniture, but it also represented the labor and skill that women put into successful homemaking.[73]

Much more of a status symbol, but still associated with female accomplishment, was the piano. The inventories bear out the conventional wisdom that the piano became an essential furnishing in nineteenth-century bourgeois salons. Whereas four households had pianos in the sample from the Restoration, the number rose to thirteen in the Second Empire sample. Only one other musical instrument appeared in the sample—a harp from the Restoration period. Music, like sewing, was considered highly appropriate for girls, though piano playing was less gender specific than sewing. Music could provide entertainment for families and friends, constituting an important aspect of bourgeois leisure. Piano playing could also be part of courtship, demonstrating female accomplishment and promising cultivation and refinement in a marriage. A piano in the salon could be a status symbol—a fairly costly item of furniture implying that the owners enjoyed leisure and knew how to use it for the pursuit of culture. Whatever the reasons for possessing a piano, demand was

72. Only one sewing machine appeared in the sample; it belonged to the wealthy Sudry family under the Second Empire. The fact that the Sudrys lived in the most expensively furnished household in the entire sample and the fact that they were the only family with a sewing machine raise some interesting questions. Were sewing machines rather expensive in the 1860s, and hence affordable only by the rich or by entrepreneurs? Did the wives of even successful professionals make their own clothes and do other sewing for the family? In short, was the sewing machine a status symbol, or was it a time-saver for women who themselves performed this particular housekeeping task?

73. J.-A. Brisset, "La Ménagère parisienne," in *Les Français peints par eux-mêmes* (Paris: J. Philippart, n.d.) 12:93.

high enough in the second half of the nineteenth century to maintain several piano manufacturers in Paris alone.

Music, card playing, and to a lesser extent fancy needlework were as much social as leisure activities. A much more ritualized aspect of bourgeois life that was also partly a form of leisure was gift giving on holidays and at certain stages in a person's life. All of these activities encouraged consumption (buying pianos, sewing tables, and card tables), but none more so than gift giving. Occasions for gift giving were many and various and included birthdays or name days, engagements, christenings, New Year's Day, Christmas, and returning home after a trip or a long absence.[74] New Year's Day was the biggest gift-giving occasion in nineteenth-century France, and gifts presented on this day had a special name—*les étrennes.*[75] Whereas Christmas gifts were primarily given to children, New Year's gifts were given to adults as well, and were not limited to the immediate family. In 1834 the author of a popular etiquette book counseled that gifts should be luxurious and elegant rather than purely useful.[76] But a manners expert in 1883 indicated that useful gifts were appropriate for social inferiors like servants and children, and for relatives.[77] Gifts that might be exchanged on New Year's Day were ribbons, books, handkerchiefs, fans, toys, trinkets, hand-sewn articles, art work, or a variety of boxes—glove boxes, handkerchief boxes, match boxes, brush boxes, and so on. Many of these articles, more elegant than useful, were, like games, products of the extensive *articles de Paris* manufacture in the capital. Some of the trinkets on dressing tables, night tables, dressers, and shelves (the assessors even had a special name for these knickknacks—*les objets d'étagère*) inventoried in the homes of richer bourgeois were doubtless gifts.

The game tables, sewing tables, pianos, and decorative objects in bourgeois households suggest a few things about leisure, status, and social ritual among members of the middle class, and their relation-

---

74. Mme Celnart [Élisabeth Félicie Bayle-Mouillard], *Manuel complet de la bonne compagnie* (Paris: Roret, 1834), 235–39. Martin-Fugier, "Rites," 260.

75. Perrot asserts that more étrennes were given in the nineteenth century than now. *Mode de vie,* 98. A letter from Madame Jullien to her husband in 1817 recounts the leading role of women in gift giving at New Year's. Archives Nationales, Fonds privés, 39 AP 4, Papiers Jullien.

76. Celnart, *Manuel complet,* 236.

77. Louise d'Alq, *Le Savoir-vivre dans toutes les circonstances de la vie* (Paris: Bureaux des causeries familières, 1883) 1:115–20.

ship to consumption. Leisure itself in the nineteenth century was a symbol of status and distinction, for only the wealthy (and increasingly the comfortable) could afford to spend time engaged in nonremunerative activities and could afford to spend money on nonessential goods. Even if family members did not play cards, never learned music, or eschewed sewing, the objects associated with these activities connoted a fairly high level of financial security, and an understanding that bourgeois men and women *should* have the leisure and cultivation to play games, make music, or sew. These activities could also serve to reinforce family ties and establish or maintain social relations with other bourgeois. Gift giving in particular required a knowledge of social hierarchy and of the gifts appropriate at each level—knowledge generally expected more from women than men.

None of the leisure objects inventoried in the sample were new products of an industrial age—as, say, bicycles or automobiles would be a few decades later. And the processes of manufacturing small tables, pianos, and *articles de Paris* reflected a demand for goods whose sole reason for existing was to manifest consumers' status and taste. The way that producers accommodated the growing demand for articles of leisure or gifts, notably in the *articles de Paris* industries, was to parcelize hand manufacturing among many different workers and tasks. This lowered the cost of production compared to having a few artisans perform all phases of manufacturing in one well-equipped shop, and it also allowed producers to change designs, colors, and forms for each new fashion season.[78] Manufacturers of tables, pianos, and *articles de Paris* appealed to consumers primarily through reputation and through innovations in style, a process that will be explained further in chapter 4.

CULTURE IN THE HOME

Two forms of bourgeois leisure that connote cultivation rather than sociability are art and book collecting. Few of the householders included in the sample can be considered true collectors of art or books, with the possible exception of Lazare Gulliet, who was head of the administration of imperial furniture, and François Abel Catherine (*dit* Lefevre) whose income came from a partnership in a

78. Alain Faure, "Petit Atelier et modernisme économique: La Production en miettes au XIXe siècle," *Histoire, économie et société* 4 (1986): 531–57.

decorative porcelain business. In the inventories of both of these households, art and art objects were listed separately; in all other cases, prints and statuettes were included with the other furnishings in each room. Judging from the occupations of Gulliet and Catherine, art appreciation was as much a part of their jobs as a hobby. The families in the sample simply were not rich enough for collecting on the scale of aristocrats or successful bankers or entrepreneurs. The prints, paintings, and statuettes inventoried were obviously intended for decoration or use; in the households of two architects and one designer from the Second Empire there were numerous drawings, doubtless related to work, as well as other works of art. Similarly, one of the most expensive book collections belonged to the doctor Charles Sudry and consisted largely of medical books and journals. It seems safe to assume that the libraries of the householders in the sample were either for professional or pleasure reading, and possibly for status; not for the rarity or beauty of the books.

For all of these reasons, and because of frequent limitations in the detail of the inventories, it is difficult to assess exactly what were bourgeois tastes in art and literature. And it is impossible to do an in-depth analysis and taxonomy of the artists and writers, genres and styles that appealed to Parisian bourgeois based on this small sample. Nonetheless, the inventories reveal some interesting trends in the consumption of art and literature related to technological and cultural developments.

The following list compares the numbers of households having prints and photographs.

	Restoration	*Second Empire*
Engravings	13	11
Lithographs	2	4
Photographs	—	3

Engravings were usually framed in gilded or painted wood and protected by glass; they were hung in the salon, dining room, office, or bedroom. The subject matter of these prints tended to the literal— views of Switzerland and parts of France, portraits of French kings and military commanders, battle scenes, and some mythological and religious subjects. Prints were probably the cheapest form of art reproduction available during the Restoration; this would account for their frequent appearance in bourgeois households. Moreover, engravings lent themselves to representations of what was familiar and clearly identifiable. For example, many men in the Restoration period had first-hand knowledge of the great Napoleonic battles, and Louis XVI had reigned fairly recently.[79] Engravings may also have reflected the purchasers' political allegiances and ideals at a time when few ordinary Frenchmen were able directly to influence the government and its leadership.

After the Restoration period, new technologies in the reproduction of art greatly expanded the possibilities for home decoration and, more debatably, for art appreciation. One great advance in the technique of two-dimensional representations was lithography. However, this was used more in the ephemeral press than for framed scenes; lithographs and woodcut prints in periodical literature brought art into more bourgeois homes than the purchase of framable prints did. Assessors rarely listed the subjects of engravings and lithographs in the Second Empire sample, so it is impossible to compare those households' taste in subject matter with that of the Restoration households.

Another new development in two-dimensional representations was photography. Three households in the Second Empire sample had photographs, and it is probable that these were images of family members. Here was yet another outlet for bourgeois consumerism— family portraits and mementos that were far cheaper than painted portraits or miniatures.

The following list compares the numbers of households that owned paintings and drawings.

	Restoration	*Second Empire*
Oil paintings	5	11
Watercolors	2	5
Drawings	3	5

The dramatic increase in consumption of two-dimensional art is likely the result of social and cultural developments (though it should

79. "[A]necdotal interest and romantic subjects drawn from history had more meaning for [the bourgeois public under the July Monarchy] than the aesthetic values of a work of art." Albert Boime, *The Academy and French Painting in the Nineteenth Century* (New Haven: Yale University Press, 1986), 14.

be noted that two architects and a designer were included in the Second Empire sample). More wealth and more education tend to increase people's interest in art and make their appreciation of it more abstract.[80] Wealth and education alike were increasing among Parisian bourgeois in the course of the nineteenth century, for both men and women.

Though the Louvre museum had been open to the public since the time of the French Revolution, the public display of the art chosen by the academy—the annual salon—started only in 1831. According to Albert Boime, the opening of the salon greatly increased public appreciation of art and the public influence on state art policies. The salons were extremely popular social and cultural events, and newspapers reported on them extensively.[81] A review of the salon of 1863 signaled critical and popular appreciation for sketches and landscapes as more innovative and spontaneous than the highly polished, historical paintings favored by the academy.[82] This shift in taste may explain the many landscape and seascape paintings on the walls of bourgeois homes in the sample from the Second Empire. As urbanization speedily progressed, consumers sought representations of untamed nature that they could enjoy within the confines of domestic safety. But if historical and mythological subjects were yielding to landscapes in the prints and paintings bourgeois householders bought, they increasingly populated interiors in the form of bronze statuettes and other three-dimensional figures under the Second Empire.

The following list compares the numbers of households possessing three-dimensional art objects.

	Restoration	Second Empire
Bronze figures on mantel clocks	7	13
Bronze statuettes	0	8
China figurines	3	5
Copper statues	1	0

80. Bourdieu, *Distinction*.
81. Boime, *Academy*, 14.
82. Ernest Chesneau, *L'Art et les artistes modernes en France et en Angleterre* (Paris: Didier, 1864), 232–33.

The bronze figures on mantel clocks represented such subjects as Hercules, Orpheus, an Amazon, Europe, and a Christian knight teaching a Saracen to read (fig. 18).

Developments in the bronze industry allowed bourgeois consumers in the Second Empire period to embellish their homes with expensive and beautiful reproductions of sculpture (see chapter 5). Art historians have remarked upon the nineteenth-century bourgeois penchant for statuary over painting, attributing this preference to their philistinism. According to this view, bourgeois men and women more easily appreciated the accuracy of three-dimensional representations as opposed to the necessarily more abstract two-dimensional images of three-dimensional subjects.[83] Moreover, sculpture was more closely related to industry, in that skilled workers executed a design created by the artist. Indeed, whereas the organizers of the Crystal Palace exhibition excluded paintings "as works of art," they deemed sculpture (among other art forms) to be admissible to an industrial exhibition because it was "connected with mechanical processes, which related to working in metals, wood, or marble."[84] Because of the "mechanical processes" involved in making statues, the reproduction of sculptures, especially on a smaller scale, was fairly easy; certainly there was no equivalent technique for painting in the middle of the nineteenth century. Thus bourgeois consumers could satisfy their desires for art more easily and more cheaply than during the Restoration, by buying bronze statuettes and mantel clocks with bronze figures representing Italian statues, mythical persons, and sentimental cherubs. With the new art reproduction techniques manufacturers unleashed a flood of demand for art in the home, and the bronze industry in particular was at its peak during the Second Empire.

The possession of art was partly an effort to imitate the aristocracy; noble patronage of the arts dated back several centuries. However, a taste for literature, though still associated with leisure-class status, also connotes other characteristics such as ambition, professionalism, and sheer personal enjoyment. The nineteenth-century Parisian bourgeoisie bought books and, increasingly, periodical literature; but the

83. Philadelphia Museum of Art, *The Second Empire, 1852–1870: Art in France under Napoleon III* (Philadelphia: Philadelphia Museum of Art, 1978), 15.

84. Quoted in Nikolaus Pevsner, *High Victorian Design: A Study of the Exhibits of 1851* (London: Architectual Press, 1951), 119–21.

inventories do not reveal whether the books were read, or exactly which titles householders owned. Families headed by professionals tended to have the most valuable book collections in the sample, but even petty entrepreneurs and property owners possessed works by great French authors. Books were owned by twelve out of sixteen householders in the Restoration sample, and ten out of sixteen in the sample from the Second Empire. This distribution is surprising, given that books were more readily available and their prices were declining after the Restoration period.[85] But the middle of the nineteenth century was also the time when periodical literature expanded dramatically, in response to an avid audience of bourgeois readers.

Reading matter listed in the inventories from the Restoration consisted entirely of books, especially volumes by French writers of the seventeenth and eighteenth centuries. Works by Voltaire appeared in seven of the twelve book collections in the sample. The next most popular author was Molière, owned by six householders; and Rousseau and Racine were noted in five inventories each. Works of history, especially the history of France, were also common.

Balzac clearly understood the tastes of bourgeois consumers under the Restoration; his inventory of César Birotteau's library could easily have appeared in the Minutier Central. Balzac also suggests that book ownership by the petty bourgeoisie was for show only. In the novel *César Birotteau,* set in the Restoration, family members buy gifts for one another on the occasion of the remodeling of the apartment above the perfume shop and Birotteau's induction into the Legion of Honor. Told by the architect that her father will soon have built-in bookcases in his room, young Césarine immediately "flung all her girlish savings upon the counter of a bookseller's shop, and obtained in return, Bossuet, Racine, Voltaire, Jean-Jacques Rousseau, Montesquieu, Molière, Buffon, Fénélon, Delille, Bernardin de Saint-Pierre, La Fontaine, Corneille, Pascal, La Harpe—in short, the whole array of home book collections to be found everywhere, which assuredly her father would never read."[86] Apparently the great men of French letters under the Old Regime were standard fare for a

---

85. Alain Corbin, "Le Secret de l'individu," in Perrot et al., *De la Révolution,* 489; Theodore Zeldin, *Taste and Corruption,* vol. 4 of *France, 1848–1945* (Oxford: Oxford University Press, 1980), 4.

86. Balzac, *César Birotteau,* 313.

bourgeoisie without a culture of its own. But by the Second Empire it would have been harder for Balzac to come up with such a standard list for a bourgeois library.

The sample from the Second Empire indicates that some book owners had works by Voltaire, Rousseau, Racine, Corneille, and Molière; but they also had books by nineteenth-century writers such as Lamartine and Victor Hugo. However, the big difference between libraries inventoried in the Restoration and in the Second Empire was that the latter frequently included issues (often bound) of periodicals — *Le Journal, L'Illustration, Revue des deux mondes, Magasin pittoresque,* to name a few. The architects Rateau and Trilhe and the medical doctor Sudry owned copies of journals relevant to their respective professions — something not found in the sample from the Restoration. Their libraries also included design and architecture books, and several volumes on medicine, respectively. In these cases, owning books and periodicals was professionally useful, and not just a matter of show as with Balzac's perfumer Birotteau.

Periodical literature served other functions as well. It kept readers abreast of domestic political developments, foreign affairs, art and literature, the social life of the elite, and human-interest stories. It also promoted consumption in a far more direct way than books, visiting, or travel could do — through advertising and through pronouncements on and analyses of taste, furnishings, and fashion. Articles like those reporting on the Crystal Palace exhibition suggested to readers what styles were fashionable, what manufacturers reputable, and what objects were indispensable to domestic happiness and social confidence. Thus leisure begat consumption, and consumption begat more consumption. Though still influenced by the aristocratic model of tastefulness, the bourgeoisie was in the process of becoming a consuming class with its own distinctive tastes and means of disseminating them.

CONCLUSION

Across three generations of bourgeois consumers, certain kinds of furnishings continued to connote high social status and were therefore considered tasteful: dark, hardwood furniture, artistic mantel clocks, thick and colorful drapery, shining candlesticks and silverware, and an open hearth surrounded by gilded or bronzed metal accoutrements. Householders from the Restoration and the Second Empire upheld a standard of tastefulness in home furnishings that in many ways resembled that of the Old Regime aristocracy. But within this framework of good taste, several factors contributed to changes in domestic interiors that both promoted an aristocratic model of tastefulness and began to delineate a more distinctly bourgeois character.

With the increasing fortunes of the bourgeoisie, the spread of education, the rebuilding of Paris, the designation of women as consumers, and technological advances in textile manufacturing and certain art industries (among others), middle-class householders of the Second Empire tended to furnish their homes with richer and more varied fabrics, more art work, more gilded candelabras, and more furniture in general than their Restoration forebears. Plentiful fabric, works of art, and a host of tableware and decorative objects from the goldsmith and the silversmith were all traditional symbols of elite status; yet many nineteenth-century bourgeois consumers could acquire such possessions only because of the increasing output in the textile, bronze, gold- and silversmith industries due to steam power, mechanization, and electrolytic metal plating. In these instances, technological developments allowed bourgeois householders to purchase aristocratic trappings at more affordable prices than the handmade articles of earlier times. Industrial innovations did not so much stimulate new tastes as permit the fulfillment of old tastes.

Where bourgeois consumers differed from their aristocratic predecessors was in the concern for domestic comfort. The inventories reveal that once the taste for status and elegance was satisfied—as with big wooden beds—then householders indulged their desire for comfort—as in buying box-spring mattresses. Yet not all such innovations of modern industry were enthusiastically embraced; the metal stove for home heating is a case in point. Tastes associated with status still prevailed over the goal of comfort, or rather, taste makers defined domestic comfort largely in terms of elegance, art, and harmony. Bourgeois consumers welcomed new products that enhanced their comfort, as long as they did not interfere with the appearance of status and wealth.

Another way in which bourgeois consumers were slowly transforming the aristocratic model of tastefulness was in the emphasis on female dress compared to that of men. Tastefulness was increasingly associated with women in the nineteenth century, as bourgeois men were primarily engaged in various forms of income earning. The inventories suggest a significant shift in spending priorities; the value of the household's women's clothes and jewelry increased while that of men's clothes, table silver, and household linen declined. Along with a well-appointed apartment, a well-dressed woman became a symbol of family status. It is important to note here the active participation of women, not only in dressing elegantly but also in tastefully furnishing the home. The metaphor of separate spheres should not be taken too literally, to mean that women were sequestered in the home and isolated from all that passed beyond its walls. The developments in consumer practices noted above would never have occurred without women's contribution. Women influenced, if they did not determine, the purchase of goods for status, comfort, and female adornment; and their tastes were not untouched by the social and economic changes occurring during their lifetime.

Social conditions, economic growth, and gender relations affected the tastes and habits of bourgeois consumers in nineteenth-century France. In just four decades it is possible to see how consumption was slowly evolving with these other developments. Consumption served to structure society, but it also changed with society. Another factor in altering consumption, and one only hinted at in this chapter, was technological change in industrial manufacturing. The difficult question for scholars has always been to determine precisely the relationship between consumption and production; did demand, or potential demand, stimulate innovations in production, or did the appearance of new and cheaper products create a demand for them? Having established the social, economic, and gender bases of bourgeois consumer taste in mid-nineteenth-century France, it is now time to relate them directly to the history of manufacturing in selected consumer goods industries, to address more fully the question of the relationship between consumption and production.

Part Two

The Effect of Bourgeois Demand on French Manufacturing

Chapter Four

The Success of Hand Manufacturing in Consumer Goods Industries

In his report on cabinetmaking at the Crystal Palace exhibition, Louis Wolowski, professor of industrial law at the Conservatoire des Arts et Métiers, asserted that "luxury . . . is the soul of progress." By luxury Wolowski did not mean the primitive extravagance or abundance characteristic of earlier societies that simply satiated basic drives rather than exciting more abstract sensibilities. Luxury for Wolowski was the alliance of art and everyday objects or habits, reflecting a civilized and spiritual society. Naturally, he considered France in the nineteenth century to be highly civilized and spiritual, and for this condition he credited the progress that manufacturers had effected, "know[ing] so well how to unite the useful and the agreeable in giving to the most vulgar objects the felicitous imprint of an advanced culture."[1] According to Wolowski, producers created a desirable form of luxury with their knowledge of art applied to the manufacture of articles of furnishing and clothing. This particular notion of progress in industry nicely complemented the bourgeois standard of taste analyzed above. It also suggested a pattern of industrial development very different from "progress" toward maximum efficiency and output.

To Wolowski, as well as to other observers at the Crystal Palace, it was obvious that producers succeeded best at imbuing consumer goods with art when they worked directly upon raw materials with their hands or with hand-held tools. Michel Chevalier and others

---

1. Commission française sur l'Industrie des Nations. *Exposition universelle de 1851: Travaux de la Commission française sur l'Industrie des Nations* (Paris: Imprimerie impériale, 1855) 7:5–7.

even suggested that art and beauty in manufactured goods diminished proportionately with technological advances in production; they referred to the exquisite workmanship and gorgeous design of Indian textiles, made entirely by hand, compared with the less fine, artistically inferior products of modern English industry.[2] Certainly these men considered European civilization superior to those societies where poorly paid hand labor was overly abundant and the sole means of industrial production, but they acknowledged that hand manufacturing was better than mechanized processes in satisfying high standards of art, quality, and tastefulness in consumer goods. The skill and talent of French workers, they asserted, were the main reasons for French success at the exhibition, implying that workers' hands—and not machines—confirmed France's reputation for tasteful and beautiful manufactured products.

Historians have long agreed that hand manufacturing in homes and in workshops was the dominant mode of production in nineteenth-century France; but only recently have they begun to study the structure and organization of hand manufacturing, and only a few have examined the relationship between this kind of production and the demands of consumers.[3] This chapter analyzes manufacturing

2. Michel Chevalier, *L'Exposition universelle de Londres* (Paris: Mathias, 1851), 9. See also *Le National*, 29 August 1851, 1; and *Journal des économistes* 29 (1851): 41–42.

3. T. J. Markovitch, "Le Revenu industriel et artisanal sous la Monarchie de Juillet et le Second Empire," *Économies et sociétés*, série AF-8 (April 1967); David S. Landes, *The Unbound Prometheus* (London: Cambridge University Press, 1969); Maurice Lévy-Leboyer, *Les Banques européennes et l'industrialisation internationale dans la première moitié du XIXe siècle* (Paris; Presses universitaires de France, 1964); Patrick O'Brien and Caglar Keyder, *Economic Growth in Britain and France, 1780–1914: Two Paths to the Twentieth Century* (London: George Allen and Unwin, 1976). Scholars who have addressed the structures and varieties of hand manufacturing in France include Alain Cottereau, "The Distinctiveness of Working-Class Cultures in France, 1848–1900," in *Working-Class Formation: Nineteenth-Century Patterns in Western Europe and the United States*, ed. Ira Katznelson and Aristide R. Zolberg (Princeton: Princeton University Press, 1986), 111–54; Ronald Aminzade, "Reinterpreting Capitalist Industrialization: A Study of Nineteenth-Century France," in *Work in France: Representations, Meaning, Organization, and Practice*, ed. Steven Laurence Kaplan and Cynthia J. Koepp (Ithaca, N.Y.: Cornell University Press, 1986), 393–417; Alain Faure, "Petit Atelier et modernisme économique: La Production en miettes au XIXe siècle," *Histoire, économie et société* 4 (1986): 531–57. Two works that link consumer demand with developments in production are George J. Sheridan, "Household

processes and developments in three different consumer goods industries to show that hand methods best accommodated the bourgeois standards of taste and style explained in chapters 1 and 2. Such an approach argues for the consideration of consumer taste as a factor promoting hand manufacturing in France at a time when mechanization, concentration, and expansion of scale were supposedly being touted as the wave of the future at the Crystal Palace. This is not to suggest that French manufacturing was backward or retarded compared to the mass production methods gaining ground in England.[4] To the contrary, manufacturers in France were often adept at organizing labor and utilizing skills in ways that satisfied a growing demand in France and abroad for artistic and stylish articles of furnishing and clothing. Such entrepreneurs were not failures for ignoring or avoiding the factory system; they were successes at mobilizing skilled and semiskilled hand workers without the capital expense and the rigidity of mass production methods.[5]

The industries examined in this chapter are wallpaper making, cabinetmaking, and production of *articles de Paris,* especially fans. These industries were selected for an analysis of the relationship between consumption and production for the following reasons. First, in all three cases the items produced — wallpaper, furniture, and fans — were decorative as well as useful; thus, consumer taste was a more important consideration for manufacturers of these articles than for producers of capital goods, or of other more functional consumer goods such as cooking utensils. Second, the French jury members' reports on these industries are particularly extensive and provide much of the information required for linking demand and

---

and Craft in an Industrializing Economy: The Case of the Silk Weavers of Lyons," in *Consciousness and Class Experience in Nineteenth-Century Europe,* ed. John M. Merriman (New York: Holmes and Meier Publishers, 1979), 107–28; Henriette Vanier, *La Mode et ses métiers: Frivolités et luttes des classes, 1830–1870* (Paris: Armand Colin, 1960).

4. Landes, *Unbound Prometheus;* Rondo E. Cameron, *France and the Economic Development of Europe, 1800–1914* (Princeton: Princeton University Press, 1961); Alexander Gerschenkron, *Economic Backwardness in Historical Perspective* (Cambridge, Mass.: Belknap Press, 1962); Richard Roehl, "French Industrialization: A Reconsideration," *Explorations in Economic History* 13 (1976): 233–81.

5. Faure, "Petit Atelier"; Charles Sabel and Jonathan Zeitlin, "Historical Alternatives to Mass Production: Politics, Markets and Technology in Nineteenth-Century Industrialization," *Past and Present* 108 (1985): 133–76.

production.[6] Finally, these industries have by and large escaped the notice of historians, despite their importance in the mid-nineteenth-century French (and especially Parisian) economy. Scholars have studied extensively the textile manufacturing, mining, and metallurgical industries, but the so-called luxury industries have not received as much attention.[7] The problem stems in part from a lack of primary source material; most consumer goods manufactures were small establishments that left fewer records than the prosperous, influential textile and metallurgical dynasties. And in part the emphasis on large-scale, mechanized manufacturing derives from the belief that this was the foundation of modern industrial capitalism and therefore is more deserving of examination than traditional and "superfluous" industries catering to a market assumed to be restricted and wealthy. But as historians revise their views of the process of industrialization in nineteenth-century France, acknowledging the importance of dispersed and hand manufacturing, the exhibition's highlighting of consumer goods production becomes very relevant.

The exhibition reports and other published sources from the middle of the nineteenth century present certain difficulties for analysis. First, these materials lack any account of relations between employers and workers; they focus almost entirely upon entrepreneurs and their actions, independent of any input or response by labor. Second, the sources of the statements about consumption are unknown. It is

6. The reports generally follow the same pattern, presenting historical background, current developments, and international comparisons for each industry or category of industry; they differ in length and focus according to the author, the type of industry, and the amount or success of French exhibits in a particular industry. For example, there is almost no mention of the dressmaking and tailoring industries in the exhibition reports because there were no French exhibitors from these sectors, despite the fact that these were thriving hand manufactures in Paris. However, the feminine press at the time of the exhibition and a few excellent secondary sources will fill this gap.

7. To name only a few works on nineteenth-century textile, mining, and metallurgical industries: Gay L. Gullickson, *Spinners and Weavers of Auffay* (New York: Cambridge University Press, 1986); William M. Reddy, *The Rise of Market Culture* (New York: Cambridge University Press, 1984); Claude Fohlen, *L'Industrie textile au temps du Second Empire* (Paris: Plon, 1956); Roland Trempé, *Les Mineurs de Carmaux, 1848–1914* (Paris: Les Éditions ouvrières, 1971); Michael P. Hanagan, *The Logic of Solidarity* (Urbana: University of Illinois Press, 1980); Bertrand Gille, *Recherches sur la formation de la grande entreprise capitaliste, 1815–1848* (Paris: SEVPEN, 1959); Arthur Louis Dunham, *The Industrial Revolution in France, 1815–1848* (New York: Exposition Press, 1955).

doubtful that systematic records of consumption for any manufactured goods (aside from food) exist, for early statisticians of industrialization in France concentrated primarily on production factors.[8] Presumably the authors cited in this chapter based their assertions about consumer tastes on their own experience as bourgeois consumers, or on inferences derived from the design and style of manufactured goods. One even senses a frustration on the part of these writers that consumption was not as "rational" or comprehensible as production. Armand Audiganne especially, in his account of French cabinetmaking in 1855, was at a loss to justify the popular appeal of highly ornamented, gigantic pieces of furniture, given the modest incomes and small apartments of most French consumers.[9]

Not even Audiganne, so consistently cognizant of the influence of consumer tastes upon manufacturing, could rest easy with the fact that consumer preference for elegant, rich-looking furnishings often contradicted assumptions about "progress" toward more efficient production and falling costs of manufactured goods. The following examination of wallpaper making, cabinetmaking, and fan making avoids this discrepancy by presenting manufacturing developments in these industries from a demand perspective. It assumes a market dominated by the tastes explained in chapters 1 and 2, and it analyzes production methods in the different industries in terms of how they accommodated bourgeois demand. From this perspective dispersed structures and hand production cease to appear retarded or backward; instead they seem logical and appropriate.

WALLPAPER MAKING

The key to understanding wallpaper making in France is its origin in the imitation of tapestry and other patterned fabrics. In the early modern period those persons who could afford to cover their walls used tapestries, wooden wainscotting, or tooled leather to provide both warmth and decoration. At that time there was no discernible interest in France for alternative forms of wall covering, though in

8. Joan W. Scott, "Statistical Representations of Work: The Politics of the Chamber of Commerce's *Statistique de l'Industrie à Paris, 1847–48,*" in *Work in France,* ed. Kaplan and Koepp, 335–63.

9. Armand Audiganne, *L'Industrie contemporaine, ses caractères et ses progrès chez les différents peuples du monde* (Paris: Capelle, 1856), 117–18.

other countries attempts at wallpaper making began in the sixteenth century. Textile workers made crude versions of flocked paper (imitating the texture of cloth) by dusting chopped wool onto paper with areas of imprinted adhesives. At the same time card and print makers produced repeated patterns by stenciling onto papers that could then be joined together to form a larger sheet. But these early forms of "wallpaper" were too small to cover walls; they usually served to line furniture or to decorate other limited surfaces. Wolowski, who was the French jury member sent to the Crystal Palace exhibition to report on wallpaper, claimed that the first wallpaper maker in France was Lefrançoise of Rouen, who in 1620 produced patterns on paper that imitated the landscapes and historical scenes woven into tapestries. The quality of these early products was undoubtedly poor, and demand for them was limited. Several improvements in the late seventeenth century, such as the substitution of woodblock printing for stencils and better methods of linking up designs on separate sheets of paper, led to the expanded use of wallpaper as a wall covering in the countryside; but consumers interested in wallpaper were more likely to choose imported items from the Far East, which far surpassed French wallpaper in quality of design and color.[10]

The real breakthrough in French demand for wallpaper occurred in the middle of the eighteenth century when English manufacturers succeeded in producing flocked papers that more closely resembled the textures, colors, and designs of fabrics, and printed papers with distemper colors that were thicker and more solid and luminous than earlier types. It was at this time, too, that the French term for wallpaper, *papier peint*, first came into use, the expression *papier de tapisserie* having been common before 1765.[11] Wealthy persons, even members of the royal court, found these new products, along with wallpapers from the Far East, suitable alternatives to actual fabrics, wood, and leather for wall coverings; this acceptance contributed to the important innovations in French wallpaper manufacturing that eventually put it at the head of European production for its design, color, and finish. The main figure in this development was the Parisian manufacturer Jean-Baptiste Réveillon.

10. Commission française, *Exposition* 7:6–7; Odile Nouvel, *Wallpapers of France, 1800–1850*, trans. Margaret Timmers (New York: Rizzoli, 1981), 8–9.
11. Henri Clouzot and Charles Follot, *Histoire du papier peint en France* (Paris: C. Moreau, 1935), 14.

Réveillon started as a hanger of imported flocked papers from England; but once established as a manufacturer he became the first to integrate many different aspects of wallpaper production into one large operation. His innovations improved the quality of paper and increased efficiency of production: he produced his own paper, assembling the sheets into rolls to speed up production; he used fast colors and superimposed them in printing to achieve new subtlety in color shades; he commissioned the best artists in different areas of the decorative arts, many of them designers for textile manufactures and for Gobelins tapestry works; he improved wallpaper hanging techniques to integrate wallpaper into the decoration and architecture of the large houses and chateaux of his customers. Though Réveillon still employed only hand printing techniques, his establishment was a huge operation at the end of the eighteenth century. Hundreds of workers each performed different tasks necessary for the production of both cheap, single-color wallpaper and expensive, multicolor products that required as many as eighty different printings from as many carved wooden plates and that imitated paintings and the finest velvets, satins, brocades, and tapestries.[12] Réveillon's tremendous success, however, ended abruptly in 1789 in the midst of worker protests that marked the beginning of the French Revolution.

Pierre Jacquemart and Eugène Bénard succeeded Réveillon and, along with other Parisian wallpaper manufacturers, upheld the capital's reputation for quality wallpaper.[13] The wallpaper industry was one of the few so-called luxury industries patronized by the state during the revolutionary period; different governments needed new wall coverings with emblems of equality and other new symbols to furnish public buildings. The imitation of fabrics, however, continued to be the staple of Parisian wallpaper manufactures. Popular designs during the Revolution, Empire, and early Restoration periods included embroidered muslin, floral patterns, drapery, and hunting scenes. A new style began at this time, following the display of the first panorama in Edinburgh in 1788; but in order to reproduce panoramic views in wallpaper a change in production was necessary.

12. Nouvel, *Wallpapers*, 10–11.
13. Commission française, *Exposition* 7:8; E. A. Entwisle, *French Scenic Wallpapers, 1800–1860* (Leigh-on-Sea: F. Lewis, 1972); Adolphe Blanqui, *Histoire de l'Exposition des produits de l'industrie française en 1827* (Paris; Renard, 1827), 288; Clouzot and Follot, *Histoire*, 122–36.

Hitherto manufacturers had imprinted a design onto paper with the aid of a sledge hammer, but early in the nineteenth century a long lever became the preferred method of exerting pressure on the carved plate. This innovation greatly improved the quality of the printed design and allowed for the vast, detailed, panoramic wallpapers that soon became popular with consumers.[14]

Until early in the nineteenth century, Paris was the undisputed center of wallpaper production in France. But a rival manufacture in Rixheim, Alsace, won national recognition under the leadership of Jean Zuber. Zuber figured prominently in the official French report of wallpapers displayed at the Crystal Palace exhibition for several reasons, not least for his innovations to improve the quality and increase the output of wallpaper in France. In addition, Zuber was one of the few prominent French manufacturers outspokenly to support free trade as a national economic policy; since Wolowski was also committed to this cause, his report lavished praise on Zuber as an exemplar of progressive and competitive French entrepreneurship. Moreover, because Zuber died unexpectedly in 1853, before Wolowski had completed his report, Wolowski did not hesitate to make his account of wallpaper at the exhibition something of a eulogy to the deceased manufacturer and his contributions to French industry. Thus the official report on wallpaper making in France was decidedly lopsided in favor of Zuber, at the expense of other larger and even more reputable firms. Nonetheless, Zuber was a major figure in the history of French wallpaper making, and his case is highly instructive on the influence of bourgeois tastes and consumer demand upon manufacturing techniques.

Under the first Jean Zuber, in the late eighteenth and very early nineteenth centuries, the Rixheim factory established a reputation for beautiful floral and landscape designs, hand printed with hand-carved wooden plates. Some of the basic steps in the block printing method of wallpaper production at this time were as follows. First, workers had to make sheets of paper, usually out of rags, and then join the sheets together end to end to form a roll about ten meters long. Grounders (*fonçeurs*) then spread the roll out on a long table, approximately the same length as the roll, to brush on the background color of the wallpaper. Grounding was a three-stage process in which three

14. Clouzot and Follot, *Histoire*, 122–65; *Rapports des délégués des ouvriers parisiens à l'Exposition de Londres en 1862* (Paris: Chabaud, 1862–64), 409.

different workers spread wet paint onto the paper with both rectangular and round brushes to form a uniform color background with no breaks. Some papers were then smoothed; workers turned the colored side facedown on the table and rolled a copper cylinder along the back to produce a smooth surface and matt coloring. Other papers were satined; in this procedure, workers sprinkled the colored side with talc and then brushed the surface, to enhance the shine of the ground color.

The block printing of designs required numerous skills and procedures. An engraver gouged the design (created by a designer) out of wooden blocks, one block for each color in the pattern. An apprentice covered the block with color from a vat alongside the press, then the printer placed the block onto the paper roll, laid another wood block or batten on top of the printing block, and with the apprentice's help put pressure on the block by means of a lever hinged to the table (fig. 19). The printer repeated this process by advancing the paper and lining up the printing block according to brass register marks on the side of the block. Any unevenness of color in the printing could be touched up by hand.

Making flocked paper involved a different procedure. Using an engraved block the printer deposited glue, instead of color, onto the paper. An apprentice sifted colored, powdered wool onto the paper and then enclosed the paper in a sort of case that formed part of the long table. The bottom of the case consisted of tightly stretched calfskin. The apprentice slipped under the table and beat the calfskin with two rods so that the powdered wool adhered to the glued portions of the paper to form the velvet effect. After the paper had been hung up to dry and any excess powder removed, the printer then used the normal printing process to superimpose lighter or darker shades on the paper.[15]

Zuber introduced the following innovations in wallpaper production in the first half of the nineteenth century. He made his own paper in continuous rolls, thus eliminating the process of pasting together separate sheets. He used several new colors, like chrome yellow, Schweinfurt green, and ultramarine. In 1826 he patented the use of engraved copper cylinders for printing, instead of wooden blocks; these cylinders could be steam powered. In 1843 he invented a device

15. Nouvel, *Wallpapers*, 20–23; *Rapports*, 407–8.

for the speedy and accurate printing of striped patterns. This consisted of a brass trough divided into as many compartments as the number of stripes desired in a wallpaper design. Small holes in the base of the trough allowed the colors in the compartments to fall through onto the paper, which the printer passed beneath the trough. This procedure produced perfectly regular stripes in many different colors; the block printing method was not only much slower but produced, even with the most skilled workers, occasional slight breaks in the stripes at the points where successive pressings met.[16] Wolowski was enthusiastic about Zuber's technological innovations, but he also cautioned readers about the limitations of such developments. He did not envision machines ever replacing the hand methods of wallpaper manufacturing in France because machines could not produce as beautiful and high-quality wallpaper.[17]

The Crystal Palace exhibition showed that technological innovations in wallpaper production were more widespread in the United States and Britain than in France. American and British manufacturers used mechanical means such as Zuber installed in his Rixheim factory, and they ran these devices with steam power (fig. 20). Wolowski's support for these developments was restrained because they produced wallpaper of inferior quality. Despite the fact that mechanical means and steam power could produce wallpaper patterns with up to twenty different colors and at much lower cost than hand printing methods, hand methods still produced wallpaper that was superior in the arrangement of the design and the permanence of the colors. Indeed, workers reporting on the exhibition of 1862 asserted

16. Commission française, *Exposition* 7:9; Nouvel, *Wallpapers,* 12–13; Musée industriel, *Description complète de l'Exposition générale des produits de l'industrie française faite en 1834* (Paris: Société polytechnique et du recueil industriel, 1836) 2:228.

17. This view is confirmed by the entry on wallpaper in the political economists' commercial dictionary published shortly after the exhibition of 1839. There the author P. Mabrun asserts that since the time of Réveillon and his contributions to the high quality of French wallpaper, "the methods of production have not changed significantly, they consist, as in cloth [printing], of applying each color in sequence by means of engraved woodblocks. Some attempts have been made recently to introduce the use of machines into this industry. Although the results have been satisfactory in some respects, we think this application will not expand much" because of the low cost of labor. Guillaumin, ed., *Dictionnaire du commerce et des marchandises* (Paris: Guillaumin, 1841) 2:1708.

that the spread of mechanization in wallpaper production after 1848 spurred on hand printers to make even more beautiful wallpaper that machines could not begin to emulate.[18] Wolowski explained the popularity of mechanization and steam power in the United States and Britain primarily by the high cost of labor in those countries. But he also noted that consumers there were less discriminating than in France, thus allowing mechanized production to be profitable. "The products [made with steam power] are rather poor in quality, but the Americans do not look at them too closely; all they want is . . . to produce large quantities quickly and cheaply."[19] By contrast, French consumers sought higher-quality wallpapers that "perfectly imitate the most varied fabrics," as one women's magazine asserted.[20]

Despite Zuber's brave forays into modern technology, Wolowski asserted that cheap labor on the continent hindered the adoption of mechanized production, and in fact Zuber's factory continued to produce mostly hand-printed wallpapers. It would be interesting to know how the more than five hundred workers in the Rixheim establishment responded to Zuber's innovations, and their role in the continuation of hand methods of production; but Wolowski ignored this topic entirely. Instead he emphasized the futility of steam power and mechanization in French wallpaper production because consumers in France did not want the goods these methods produced.

The vast field open to our wallpaper industry does not extend in that direction. Stimulated by the progress of general affluence and by the increasingly widespread appreciation of tasteful objects, the production of fine-quality wallpaper develops more and more. . . . The more discriminating habits and the quest for interior comfort increase, the more there is a demand for, and consequently the production of, good wallpaper.[21]

Was Wolowski here denying the inevitable in asserting that hand production, rather than mechanization and steam power, was the path to success for French wallpaper making? After all, according to the worker delegates' report on wallpaper at the 1862 exhibition,

18. *Rapports,* 409–10.
19. Commission française, *Exposition* 7:7.
20. *La Gazette des salons,* 10 June 1838, 505.
21. Commission française, *Exposition* 7:13–14.

steam-powered, mechanized methods of grounding, printing, and smoothing were commonly operating in France at that time.[22] Was Wolowski deluded in believing that a general rise in the standard of living of French people would automatically be accompanied by consumer tastes for beautiful, well-made wallpaper? Certainly the histories of wallpaper in particular and of French industry in general suggest that Wolowski was wrong. Historians of French wallpaper maintain that its quality peaked right before the Crystal Palace exhibition and thereafter declined steadily with the expanding use of modern technology.[23] And the conventional wisdom is that the increasing disposable income of middle-class and working-class men and women meant a rise in the demand for cheap, gaudy, and tasteless goods that machine production nicely accommodated, to the detriment of handmade articles of taste and quality.[24]

It is difficult to believe that Wolowski lacked vision regarding the transforming capacity of machine technology and steam power, for he was a professor of industrial law at an institution committed to the promotion of progress, including science and technology in industry. Nonetheless, he was a man of his time; he comprehended mechanization as merely one of several alternatives for manufacturing to meet changing demand.[25] Moreover, as a proponent of free trade Wolowski also appreciated the logic of the French wallpaper industry continuing to do what it did best—that is, produce high-quality wallpaper by hand—given the demonstration at the exhibition of Britain's and the United States's superior ability in the manufacture of

22. *Rapports*, 410–11. It is tantalizing to find in this report no complaints about the loss of jobs or a decline in the quality of wallpaper as a result of mechanization. Instead, the delegates emphasized the problems of low pay, long hours, and unhealthy air in the wallpaper industry. Ibid., 413–14.

23. Nouvel, *Wallpapers;* Entwisle, *French Scenic Wallpapers.*

24. Christopher H. Johnson, "Economic Change and Artisan Discontent: The Tailor's History, 1800–48," in *Revolution and Reaction: 1848 and the Second French Republic,* ed. Roger Price (New York: Croom Helm, 1975), 87–114; Lee Shai Weissbach, "Artisanal Responses to Artistic Decline: The Cabinetmakers of Paris in the Era of Industrialization," *Journal of Social History* 16 (1982): 67–81.

25. See Louis Bergeron, "French Industrialization in the Nineteenth Century: An Attempt to Define a National Way," *Proceedings of the Annual Meeting of the Western Society for French History* 12 (1984): 154–63; Faure, "Petit Atelier"; Neil McKendrick, "Home Demand and Economic Growth: A New View of the Role of Women and Children in the Industrial Revolution," in *Historical Perspectives: Studies in English Thought and Society,* ed. Neil McKendrick (London: Europa, 1974), 152–210.

cheap, lower-quality wallpaper with machines and steam power.[26] And indeed, at the 1862 exhibition the quality of French hand-printed wallpaper continued to surpass that of all other countries.[27] As for consumption, surely Wolowski referred to the growing middle-class, rather than the laboring poor, as the main market for fine wallpaper. His report for the French jury implied that inventions like the copper cylinder and Zuber's machine for printing stripes could adequately meet the demand for cheap wallpaper without replacing hand printing for higher-quality products.

It is important to keep in mind that the shift from a predominantly bourgeois market to a mass market in France was long and gradual, as Rosalind Williams implies when she dates the appearance of mass consumption in the last two decades of the nineteenth century.[28] And as Williams also maintains, the advent of mass consumption did not replace the bourgeois standard; rather, the bourgeois standard, instead of being the sole model of consumption, became one of several consumption types. Thus Wolowski's position that hand production of good-quality wallpaper was appropriate for French industrial success, was valid for several reasons. First, modern technology was not good enough to produce better-quality wallpapers. Second, France had a reputation for excellence in the manufacture of fine wallpaper and was able to export a fair amount of wallpaper to consumers abroad—approximately three million francs' worth in 1852 alone.[29] This was a trade that Wolowski and Zuber were certain would flourish even more when the government lifted tariffs and prohibitions on foreign products entering France. Third, the majority of consumers in mid-nineteenth-century France wanted complicated patterns and fast, brilliant colors in their wallpaper, a demand that only hand printing methods could meet.

The designs of award-winning wallpaper by Zuber and other French producers at the exhibition confirm the popularity of wallpaper patterns that successfully imitated the richest, most varied fabrics and the most intricate outdoor panoramas. Granted, these items on display were masterpieces of the wallpaper maker's art and not

26. See chapter 7.
27. *Rapports,* 412.
28. Rosalind H. Williams, *Dream Worlds: Mass Consumption in Late Nineteenth-Century France* (Berkeley: University of California Press, 1982).
29. Commission française, *Exposition* 7:16.

intended for common consumption. But this is precisely the point of the exhibition in general and French entries in particular that has been so long overlooked: commentators went to the Crystal Palace as much to see the finest examples of human manufacturing as to see the progress in industrial technology. Appreciation for tasteful, artistic, well-made consumer goods was widespread, and if few could afford to buy the masterpieces that filled the exhibition hall, consumers nonetheless viewed them as models or guidelines for other producers to approximate (usually by hand methods of manufacturing).

A good example of wallpaper is Zuber's gorgeous floral pattern representing flowers from all over the world in brilliant color and rich depth (fig. 21). The design is beautiful, and the texture of color and subject matter truly deep and varied. The object of this and other wallpaper designs was not simply to imitate nature or other manu-factured materials, but to use color in such a way as to make a papered wall a work of art in itself. Tastes and techniques had evolved to a point where consumers and producers judged wallpaper on its own terms, that is, how well it would embellish the walls of urban apartments, adding distinctive color, style, and design to a room.

Another important exhibitor of wallpaper was Étienne Delicourt, a Parisian manufacturer and the winner of a grand medal. Delicourt exhibited wallpaper representing a hunting scene with hounds killing a stag in a natural forest setting (fig. 22). Outdoor panoramas like this were extremely popular at midcentury, perhaps because they pro-vided consumers with the illusion of being close to nature when in fact their distance from it was increasing or because bourgeois house-holders enjoyed the association with aristocratic leisure activities. Such designs entailed countless hours of hand labor. Wolowski noted that this one large landscape required more than 4,000 woodblocks and cost over 40,000 francs even before the first printing. Though Delicourt employed over 3,000 workers and produced annually over 700,000 francs' worth of wallpaper, Wolowski never even hinted at mechanized procedures in the Delicourt works. Instead he indicated that Delicourt owed his success to his good taste, artistic ability, and extensive research in the designing of historical and religious themes for large, luxurious wallpaper panels.

In wallpaper manufacturing, then, consumer tastes were one of at least two significant factors (the other factor Wolowski mentioned was cheap labor) promoting hand methods of production at the expense of mechanical means and steam power. Though the techno-logical capacity for mechanized production existed in France at mid-century, block printing by hand was the main method of manufac-turing wallpaper. An examination of patents taken out for new inventions in wallpaper manufacturing between 1830 and 1860 re-veals that most were in the area of block printing rather than steam-powered, roller-print methods of production.[30] The tastes of French consumers for original, artistic, and multicolored wallpaper patterns apparently prevailed upon manufacturers, even innovators like Zu-ber, to continue using hand methods of block printing and flocking.

CABINETMAKING

Perhaps the most successful category of goods that France exhibited in the Crystal Palace was furniture. Of the one hundred medals and honorable mentions awarded by the international jury to furniture producers, twenty-nine went to France, including three of the four grand prize medals. Only Britain came close to France in this sweep of awards, winning twenty-seven—and there were more British entries.[31] Observers unanimously agreed that French furniture was a model of tasteful design, harmonious decoration, skilled craftsman-ship, and refined finish. Wolowski, reporting on furniture for the official French jury sent to the exhibition, attributed French success in cabinetmaking to the tradition of producing for the sake of art, style, and quality rather than profit. Wolowski claimed that this ethos started with Jean-Baptiste Colbert's organization of the Gobelins works as the royal manufacturer of furniture under Louis XIV in 1667, and that it still prevailed among Parisian furniture makers in 1851.[32] While this claim seems highly exaggerated, certain tendencies in consumer tastes for furniture suggest why Parisian cabinetmaking retained many of the qualities that distinguished it at its apogee in the eighteenth century.

As mentioned in chapter 1, old styles of furniture were very popular among French consumers, particularly styles from the eigh-teenth century and the Renaissance. The effect of this fashion on the furniture trade and on the quality of furniture was ambiguous. On the

30. Nouvel, *Wallpapers,* 24.
31. Commission française, *Exposition* 7:44.
32. Ibid., 9–10.

one hand, critics note that the lack of any distinctive style of French furniture from the end of the First Empire (1804–14) until the introduction of art nouveau at the end of the nineteenth century was a sign of decline in cabinetmaking and of lack of imagination and creativity among cabinetmakers. Worse, the fashion for old styles encouraged copying, even fraud, as some cabinetmakers tried to pass off imitations as real antiques.[33] On the other hand, the demand for old styles may well have preserved traditional cabinetmaking skills in France and contributed to the high standard of quality of French cabinetmaking at the exhibition.[34] Indeed, some art historians discern originality and vitality in the way that mid-nineteenth-century cabinetmakers took their inspiration from the best features of eighteenth-century furniture while creating entirely new pieces.[35]

Like their aristocratic predecessors, bourgeois consumers in the middle of the nineteenth century preferred their furniture made of wood that was hard, dark in color, pleasantly scented, fine-grained, and receptive to polish for a gleaming finish. In the late eighteenth century exotic woods imported from the distant continents of South America, Africa, and Asia satisfied this demand, but starting with the French Revolution several French governments encouraged the use of native woods in cabinetmaking, initially to lower the cost of furniture when sales were depressed.[36] Despite the continued concern of the July Monarchy (1830–48) to perpetuate the use of native woods—such as oak, beech, walnut, and to a lesser extent native maple—for furniture, the most popular wood for home furnishing at the time of the exhibition was mahogany.[37] By this period mahogany furniture

33. Theodore Zeldin, *Taste and Corruption,* vol. 4 of *France, 1848–1945* (Oxford: Oxford University Press, 1980), 72–82; *French Cabinetmakers of the Eighteenth Century* (New York: French and European Publications, 1965), 326–27.

34. "[The] appearance of forgotten forms, of original designs, has inspired new desires among consumers, and has enriched and purified the taste of workers. . . . . Thus, heavily distorted veneering, furniture loaded with thousands of pieces of wood, . . . stiff and heavy furniture poorly imitating the Greeks and Romans, have given way in the last twenty to twenty-five years to today's furniture, which is accommodating, slim, and elegant." Guillaumin, *Dictionnaire* 2:1502.

35. Denise Ledoux-Lebard, *Les Ébenistes parisiens du XIXe siècle, 1795–1870,* 2d ed. (Paris; F. de Nobèle, 1965), xvi–xvii.

36. Archives Nationales, Ministère de l'Agriculture et du Commerce, F^{12} 2410, Ébenisterie, papiers peints, an III–1852.

37. Paul Garenc, *L'Industrie du meuble en France* (Paris: Presses universitaires de France, 1958), 78–81; Mme Pariset, *Nouveau Manuel complet de la maîtresse de*

was affordable on a much wider scale than in earlier times for at least two reasons. First, much French furniture was veneered rather than solid. That is, for decoration or to give the appearance of greater richness, an almost paper-thin layer of beautiful, exotic wood was glued onto a piece of furniture made of another, less expensive wood.[38] The widespread use of veneering was not necessarily, though it could in some cases be, a reflection of decline in cabinet-making. Even in the eighteenth century, the heyday of cabinetmaking, veneered furniture was popular among wealthy, aristocratic consumers because it was more artistic than solid wood furniture, allowing the inlaid woodworker to use his skill and imagination in designing and constructing a decorative panel for a chest, cabinet, or wardrobe.[39] Second, consumer taste for beautiful wood required that furniture made of native wood must use the finest part of the wood, the core or bole. The cost of extracting this core in France was so high that "a piece of furniture made of mahogany was cheaper than furniture made of French wood."[40] In addition, more people than before could afford to buy furniture.

Consumer demand for beautiful wood and old styles in furniture meant that many of the skills and procedures of cabinetmaking in the eighteenth century continued into the nineteenth century. Cabinet-making actually consisted of two different but complementary skills—cabinetmaking proper, and joining. A cabinetmaker did the veneering, inlay work, and carving on fine pieces of furniture; a joiner manufactured the frames for beds, chairs, and sofas, and produced chests of drawers, wardrobes, and cupboards made of plain, solid wood. In the eighteenth century cabinetmakers cut the thin sheets of wood for inlay work with a hand saw used in conjunction with a special chair that contained a rack and vise to hold the wood while sawing.[41] A long saw, handled by two men, cut a round core of wood into slightly thicker sheets for veneering, a technique that

---

maison (Paris: Roret, 1852), 15, 18; *Les Modes parisiennes* 424, 22 March 1851, 3010; *Le Conseiller des dames,* November 1848, 6–8; Adeline Daumard, *La Bourgeoisie parisienne de 1815 à 1848* (Paris: SEVPEN, 1963), 136.

38. Guillaumin, *Dictionnaire* 1:302, 318.

39. Alexandre [André Jacob?] Roubo, fils, *L'Art du menuisier ébeniste* (Paris, 1774).

40. Guillaumin, *Dictionnaire* 1:319.

41. Jean Nicolay, *L'Art et la manière des maîtres ébenistes français au XVIIIe siècle* (Paris: Pygmalion, 1976), 781–84; *French Cabinetmakers,* 16.

continued into the Restoration period.[42] During the 1820s and 1830s cabinetmakers began using machines for these types of sawing, allowing for much thinner veneers and more precise cutting of inlay wood, especially for repeated geometric designs. One result of this innovation was bigger furniture with more veneering, no doubt satisfying to consumers concerned about appearing rich through their home furnishings.[43] By the middle of the nineteenth century steam-powered machines were also capable of sawing, molding, sculpting, and cutting wood into a wide variety of shapes, but they were far from totally supplanting hand labor in cabinetmaking, especially in the more expensive range of furniture.[44] It is not at all clear to what extent mechanization was transforming cabinetmaking by the time of the Crystal Palace exhibition, but the continued demand for old styles that required realistic inlay designs and intricate carving suggests that hand methods of manufacturing prevailed; this conclusion is supported by recent research on the condition of cabinetmakers.[45]

The short-lived liberal newspaper *Le Bien-être universel* asserted that consumer taste for elegant, highly decorated furniture inhibited the use of machines (notably molding machines) in production: "New [furniture] is decorated with lathed or sculpted ornaments, moldings, and so on, which, in spite of the machines mentioned above, give much more work to workers and require greater application from them."[46] In addition the popular press suggests that French consumers were sensitive to good craftsmanship and quality in furniture; after all, households spent a lot of money on furniture, and it was an important indicator of wealth and status for bourgeois men and women.

The advertisements in the feminine press placed by individual cabinetmakers invariably emphasized the art and quality of their products over their cost. The publicity for a special exhibition and sale of Saint-Antoine products was particularly enthusiastic because

---

42. Roubo, *Art,* plate; *Rapports,* 258.

43. *Rapports,* 259; Commission française, *Exposition* 7:39–42.

44. Commission française, *Exposition* 7:39–42; Garenc, *Industrie,* 81, 87–88.

45. *Rapports,* 259; Weissbach, "Artisanal Responses". A contemporary study of Saint-Antoine cabinetmaking indicates that hand methods still predominate among the remaining artisans in the area. Odile Luginbuhl-Hargous, "L'Ébenis-terie au Faubourg Saint-Antoine: Tradition et transformations," *Ethnologie française,* n.s. 12 (1982): 361–72.

46. *Le Bien-être universel,* 22 June 1851, 12.

the quality of the furniture was guaranteed. This sale was organized by Saint-Antoine cabinetmakers in 1850 to pay back the 400,000 francs that the government of the Second Republic loaned them in 1848 to stimulate the recovery of the industry.[47] A commission of experts inspected all pieces submitted to the exhibition and sale, with the express intention of excluding shoddy articles and enhancing the reputation of French cabinetmaking. The feminine press was delighted with this idea, and one writer explained that while consumers would be disappointed if they expected fantastically low prices at the sale, they would definitely get their money's worth:

> We are speaking of a real bargain. . . . The furniture is made with integrity, with wood that has made *its effect,* produced with the greatest care, instead of being badly glued, fitted in haste with pieces of unseasoned wood. At first, perhaps, the buyer does not find the difference in price he expected, and cannot profit from, as they say, a good deal; but true connoisseurs will quickly appreciate the difference in the type of work, and use and time will prove to the others that they really made *a bargain.*[48]

Did bourgeois consumers actually follow these precepts and consider good quality an essential ingredient in a furniture bargain? Several sources suggest that they did. Persistent in reports of French exhibitions from 1827 through 1855 is the comment that producers emphasized decoration, beauty, and style in furniture above usefulness and affordability. Though observers like Adolphe Blanqui and Armand Audiganne criticized this tendency and called upon cabinetmakers to accommodate ordinary consumers rather than producing works of art for the wealthy elite, they also indicated that consumer demand was the root cause for the predominance of highly decorated and expensive furniture. "Our cabinetmakers," wrote Blanqui in 1827, "are very attentive to the tastes of the public, and they cater to them by accommodating their tastes, or rather whims."[49] Almost thirty years later Audiganne echoed the same concern, suggesting that French cabinetmakers' exhibits would continue to be richly

47. Chambre de Commerce et d'Industrie de Paris, letter to the prefect of the Seine, 6 April 1850, in Correspondance, 9 April 1846 to 18 June 1850, p. 249.
48. *Le Foyer domestique,* 1 June 1850, 421. See also *Le Conseiller des dames,* May 1850, 220; June 1850, 250.
49. Blanqui, *Histoire de l'Exposition,* 151.

ornamented, costly, and impractically large until consumers came down to earth and matched their tastes with their incomes and the needs of small apartment living. "Give [the public] time to realize that it is impossible both to use and to pay for these magnificent pieces of furniture, and these eccentric tastes will recede rapidly."[50] Audiganne seems to have misapprehended consumer motivation, thinking that bourgeois householders would "learn" to buy furniture solely on the basis of utility and practicality. He obviously saw little merit in consumer tastes for big and highly decorated pieces, but his contemporary Wolowski took a different view of such preferences and their effects upon the quality of furniture made in France.

Far from deploring the fine and expensive products of the cabinet-makers' craft, Wolowski supported their production as beneficial to the industry. Defending the nation's cabinetmaking from the accusation that manufacturers cared too little for consumers of modest means and simple tastes, Wolowski asserted that masterpieces like those exhibited in the Crystal Palace "maintained and elevated the traditions of taste and prevented inspiration from dying out."[51] With such models, he contended, cabinetmaking in France would uphold its reputation for quality and tastefulness, even among its cheaper products. Workers, too, were happy to echo this sentiment in 1862, stating that the popularity of eclectic styles since 1830 had encouraged cabinetmakers to become more adept at decorating furniture with moldings, sculpture, copper, tin, mother-of-pearl, porcelain, and marble. They also maintained that cabinetmakers' skill in composition and fitting of furniture pieces was almost perfect in the middle of the nineteenth century.[52]

This evidence contradicts Lee Shai Weissbach's assertion that urban middle-class demand for "reasonably priced" furniture early in the nineteenth century was a form of mass consumption that stimulated highly divided, large-scale, and mechanized forms of production in cabinetmaking.[53] This interpretation fails to distinguish between "bourgeois" and "mass" consumption, and it does not acknowledge the social and cultural connotations of furniture for bourgeois consumers seeking to assert class status. If low cost were

50. Audiganne, *Industrie contemporaine,* 118.
51. Commission française, *Exposition* 7:7.
52. *Rapports,* 259.
53. Weissbach, "Artisanal Responses."

the primary concern of bourgeois consumers, then artisan cabinet-makers should have fallen much more rapidly before the onslaught of mass production techniques. Instead, skilled and semiskilled workers produced furniture that imitated older styles throughout the century; this production possibly represents, as Weissbach states, a diminution of artistic creativity and artisan control, but it also indicates that consumers valued style, status, and some degree of quality rather than low cost. Division of labor, expansion of scale, and structural reorganization of marketing and production undeniably caused a gradual decline in both the quality of furniture and the status of artisans during the nineteenth century, as Weissbach asserts;[54] but Weissbach's most persuasive evidence for such decline indicates that it was most palpable at the end of the century.

The growing market for furniture in the nineteenth century caused significant changes in the structure of the cabinetmaking industry, and interpreting these developments is difficult. One type of organization was a shop in which a manufacturer/owner employed workers and also sold the furniture made on the premises. Usually such manufacturers were accomplished cabinetmakers themselves, and contemporaries regarded the products from these shops as being among the finest examples of cabinetmaking.[55] More controversial was the practice of *façonnage,* which was by no means unique to Parisian cabinetmaking.

The *façonneur* worked at home, either alone or with a few assistants. Though possessing little capital, he had to pay for the rent of an apartment and/or work space, and for all tools, wood, and other materials necessary for cabinetmaking, to say nothing of workers' wages. Since *façonneurs* were usually paid at the same piece rate as workers in shops, the extra burden of overhead costs often meant that these independent cabinetmakers worked extremely hard, and even then could barely make ends meet. For *façonneurs* who were exceptionally lucky or skilled, the fact of independence might outweigh the difficulties of covering the extra costs, and they might earn a

---

54. The article in *Le Bien-être universel* deploring the low pay and misery of furniture workers asserted that work was *not* divided, though from the eighteenth century on, furniture making included many different trades, such as sawing, lathing, molding, sculpting, polishing, upholstery, and fancy braid. *Le Bien-être universel,* 22 June 1851, 12.

55. *Rapports,* 261; Audiganne, *Industrie contemporaine,* 113–14.

subsistence, if not prosperity, at least equal to that of wages earned in a shop. Indeed, Audiganne suggests that the majority of cabinetmaking in the faubourg Saint-Antoine after 1848 took the form of successful *façonnage*.[56] He praised highly the cabinetmaker Jeanselme, who won recognition at the Crystal Palace exhibition, for his rise from humble *façonneur,* through education, skill, hard work, and luck, to the ownership of a flourishing establishment employing some three hundred workers.[57] Nonetheless, most *façonneurs* earned a difficult living; they enjoyed a certain amount of independence, and one artisan publication touted the advantage of spending the workday with wife and children; but the only clear beneficiaries of this system were the retailers who resold at a handsome profit the furniture made by the *façonneurs.* Workers viewed this type of marketing as pernicious to cabinetmaking, since the retailer often knew nothing about the craft and was as likely to sell junk as quality furniture if he could profit from doing so.[58]

Another way of producing furniture was called *la trôle.* Initially this practice was an assertion of workers' control over their own labor. Each worker manufactured a whole piece of furniture, rather than repeatedly performing a single task as was usual in large workshops or even those run by *façonneurs.* Moreover, the *trôleur* sold directly to the customer, not to a retailer. At the time of the Crystal Palace exhibition both *façonnage* and *la trôle* were common in the cabinetmaking trade, and both represented the viability of dispersed manufacture and of hand production methods in an industrializing age. By the end of the nineteenth century, however, *la trôle* usually meant furniture that was cheap and poorly made.[59]

56. The 1848 Revolution was devastating for the *façonneurs.* Since the furniture market was so depressed at this time, *façonneurs* especially had to sell their pieces at unconscionably low prices in order to pay their own debts. This further drove down the prices of furniture. Chambre de Commerce et d'Industrie, *Statistique de l'industrie à Paris, 1847–48* (Paris, 1851), 157–60.

57. Audiganne, *Industrie contemporaine,* 109, 114. The figures gathered for the 1848 survey of Paris industry suggest how rare Jeanselme's example was: 178 cabinetmakers employed more than ten workers; 844 employed two to ten workers; 448 employed one worker; and 445 worked alone. Chambre de Commerce, *Statistique,* 157.

58. *L'Atelier,* December 1840, 29; *Rapports,* 261–62; Faure, "Petit Atelier," 549.

59. Faure, "Petit Atelier," 552–53; *Le Bien-être universel,* 22 June 1851, 11–12; Weissbach, "Artisanal Responses."

The subject of changes in furniture production in the nineteenth century and their effects on workers requires more study, but accounts of the exhibition suggest that consumer demand for solid, beautiful, and tasteful furniture encouraged the persistence of hand methods of manufacturing. Machines could not work the hard woods nor produce the intricate decoration that bourgeois women and men desired in their furniture. As long as old styles were popular and furniture retained the strong association with high social status for bourgeois consumers, handmade, fairly good-quality beds, tables, sofas, buffets, desks, and so forth were in high demand. Division of labor and the organization of production under the authority of a merchant capitalist who knew little about cabinetmaking but possessed skill in retailing diminished the quality of furniture and the status of workers, but such methods still allowed for skilled hand manufacturing and for flexibility of style. Accounts of wallpaper making, cabinetmaking, and fan making (described in the next section) suggest that division of labor had been common under the Old Regime and was not necessarily an innovation of nineteenth-century merchant manufacturers. Moreover, it is likely that *façonneurs* and *trôleurs* produced mainly by hand, since only larger operations could afford the capital outlay on machines.[60] Not only were manufacturers pushed toward hand production methods for lack of incentive toward technological innovation; consumer demand for quality and style also pulled them in this direction. French accounts of furniture at the exhibition praised and encouraged the traditional methods and characteristics of cabinetmaking that rendered French furniture so superior to that produced in other developing countries. Beauty and quality, and the hand manufacturing these entailed, were the strengths of French cabinetmaking; and reporters expressed limited enthusiasm for mechanization and its benefits to the trade and to furniture. Consumer taste, then, was among the factors behind the success of hand manufacturing in French cabinetmaking in the nineteenth century.

ARTICLES DE PARIS: FAN MAKING

An important but little studied group of industries in France was the production of *articles de Paris*—artificial flowers, combs, brushes,

60. Garenc, *Industrie du meuble,* 87.

buttons, pins, buckles, parasols, canes, fans, dolls, games, and a wide range of decorative boxes. As the name suggests, Paris was the center of the manufacture of these goods, and these trades collectively comprised one-third of all Parisian industry.[61] *Articles de Paris* were above all items of fashion, and therefore subject to constant and rapid changes in design, style, and composition. At the Crystal Palace exhibition, although manufacturers in Britain, the German states, and Belgium also exhibited accessories and knickknacks, the goods from Paris were unrivaled in their taste, elegance, and art. Parisian producers catered to highly volatile and exigent consumer tastes: "It is not enough to administer the manufacture; it is necessary, each year, to create new models, to adapt subjects, designs, ornaments, devices, and colors to the market, to the circumstances, to the current fashion, to the tendencies of the public."[62]

Natalis Rondot, a delegate of the Lyon Chamber of Commerce to the exhibition who reported on fancy articles, was torn between accolades for French superiority in taste, art, and creativity, and regret for the decline of manufacturing from an art to an industry. New machines, the decline of apprenticeships, the use of cheaper raw materials, and the division of labor were also responsible for a loss of artistic ability among workers, according to Rondot. He also implied that the contemporary expansion of the market for *articles de Paris* to include bourgeois with somewhat limited financial resources encouraged greater efficiency and lower production costs, at the expense of craftsmanship and art. Nonetheless, Rondot asserted that low prices could not be the overriding goal of manufacturers, or the goods would lose the very qualities that made them desirable to consumers.

> Trades that confined themselves to art have lost their elevated character. It is certainly a contribution to progress to provide them with the means of lowering the costs of production, and we would applaud this benefit; but it is not enough, for success is . . . less in the price than in the form. It is necessary to give to our designs, to our fashions, to our imaginative creations, a new originality, and to seek in the study of masterpieces of all times the secret of enduring elegance and new attractions.[63]

61. Commission française, *Exposition* 7:6.
62. Ibid., 31.
63. Ibid., 9.

From Rondot's perspective, what made French *articles de Paris* so tasteful and so popular with domestic and foreign markets was both their imitation of old styles and their originality—two seemingly antithetical characteristics that comprised bourgeois taste as defined in chapter 1. Though machines and division of labor could lower costs and increase output in these consumer goods industries, hand production still predominated as the means to meet the demand for fashionable knickknacks and accessories.[64] In fan making in particular, fantasy, not practicality, was what 'consumers wanted in their product: "Imagination . . . dominates the fan trade. Tastes vary infinitely. The merchant must make his principal study that of whims. . . . Appearance is what [the consumer] seeks above all; he or she will hardly deign to consider the efforts made to ensure the solidity or durability of the object that has appealed to him or her."[65] Machines aided workers in some of the early, less artistic phases of fan manufacture, and work was divided among a wide range of rural and urban men and women. Yet these developments did not eliminate high levels of skill and competence among workers or an effective, dispersed structure for accommodating bourgeois tastes. Moreover, machines did not always replace hand methods in fan making, for they were sometimes less efficient. "Since 1810 several different attempts to produce ribs by mechanical means have not succeeded." Hand workers, Rondot went on to say, were actually less wasteful of materials than machines.[66]

The making of a fan involved eighteen to twenty workers, performing different jobs in three main series of operations. The first step consisted of work on the mounting, or the foot, as fan makers called the frame and ribs which supported the screen. Approximately 1,200 men, women, and children cut, shaped, and decorated the wood, ivory, bone, or shell for fans, and almost all lived in communes near Paris in the department of the Oise. Working at home with help from other family members, a worker first sawed the ribs using a small machine fashioned with watch springs. Another worker filed the ribs to give them shape. The ribs then passed to a polisher,

64. Rondot indicates that the putting-out of hand work was usual in the manufacture of combs, brushes, toys, boxes, and fans. Commission française, *Exposition* 7:9.

65. Guillaumin, *Dictionnaire* 1:912.

66. Commission française, *Exposition* 7:82.

a jigsaw cutter, an engraver, a sculptor, a gilder, and a worker who covered them with tiny sequins of gold, silver, or other metal. These completed ribs were then pierced at the base to allow for their joining; this operation was so important that Rondot's report on the exhibition included the names of piercers noted for their skill.

The second series of operations, occurring in Parisian workshops, produced the screen. Fan screens were made of vellum, parchment, paper, taffeta, satin, crepe, or silk gauze. A designer provided the composition; either a painter or a lithographer transferred it to the screen, and a colorer added more hues. Folding the screen and gluing it to rib extensions completed the screen work. The fan then passed to a third set of artisans for the final stage, that of mounting. A mounter attached the screen to the ribs, and a decorator and a borderer added ornaments of gold or bronze, colors, and even tiny mirrors to the screen and ribs. A female worker inspected the fan at the end, attaching to it a tassel or feather and choosing a small case in which to enclose it.[67]

Orchestrating these operations was the fan manufacturer, *l'évantailliste,* in whose Paris shop only the final mounting stages occurred. In addition to supervising the mounting, the fan manufacturer gave rib designs to peasant workers, directed the screen painters or printers, chose the decorations for the ribs and secured their execution, and coordinated the diverse tasks "to obtain an original and well-made product at the lowest possible cost."[68] (See fig. 23.)

Rondot exaggerated the newness of machines and division of labor, at least in the case of fan making; a similar parcelization had existed in the eighteenth century—the apogee of art and beauty in French fans—and the machines made of watch works hardly seem on the cutting edge of modern technology. "The production of fans has always employed a large number of workers from very diverse professions—inlay, gilding, glittering mirrors, paper, plumes, painting, and embroidery—coming together in the composition of this so useless object, which, simple or decorated, rich or middling, is no less the work of several trades united in one."[69] Moreover, neither machines nor division of labor significantly diminished workers' control over their labor. Peasants themselves devised the machines they used

67. Ibid., 30–41.
68. Ibid., 82.
69. Guillaumin, *Dictionnaire* 1:910.

to cut and shape the fan ribs, and most of the work on fans was more or less independently contracted and performed in the workers' homes.[70] What the workers may have lacked in artistic training and an overall view of the intended design and shape of a fan, they made up for in superior ability in their designated tasks. Rondot quoted an *éventailliste* as saying that "the fan manufacturer of Paris provides only initiative, advice, and advance payment"; the real work was carried out by piece workers in shops and homes.[71] Rondot concluded that taste and skill on the part of the worker, rather than advances in production methods, explained French superiority in *articles de Paris*.

These methods vary as often as the models, and tools are always the simplest; it is the alert and skillful hand of the worker that alone can give to these marvelous things the delicacy that distinguishes them and the freshness that adorns them. Of the myriad ideas, inventions, improvements that have emerged from Paris for a century, the majority relate to form; the merit of manufacture is only secondary.[72]

CONCLUSION

The examples discussed in this chapter demonstrate the suitability of hand methods of manufacturing to meet bourgeois demand for tasteful and stylish consumer goods in mid-nineteenth-century France.[73] They also suggest the increasingly important role of the manufacturer as a mediator between producer and consumer. Before the expansion of bourgeois consumption, artisans often worked directly for

70. Faure, "Petit Atelier," 539–40.

71. Commission française, *Exposition* 7:97. To be sure, the fan manufacturer fulfilled the essential function of responding to different markets: "Each country has its own fashion and taste for fans, and the Parisian fan maker must vary the materials, styles, decoration, and subject matter according to the fans' destination." Ibid.

72. Ibid., 146.

73. Another good example of the viability of hand manufacture is the garment industry. Deskilling and division of labor occurred under the July Monarchy in the manufacture of men's clothing, primarily for a working-class clientele; but custom manufacture persisted for wealthier consumers and notably for women's apparel. Vanier, *Mode;* Johnson, "Economic Change"; Philippe Perrot, *Les Dessus et les dessous de la bourgeoisie* (Paris: Arthème Fayard, 1981); Whitney Walton, "'To Triumph before Feminine Taste': Bourgeois Women's Consumption and Hand Methods of Production in Mid-Nineteenth-Century Paris," *Business History Review* 60 (Winter 1986): 541–63.

wealthy, aristocratic patrons. But larger and more varied markets required a manufacturer to be ever alert to changes in taste, style, design, and color, and then to act quickly to inform workers of the necessary modifications in production. Production-oriented analyses of the rise of the manufacturer have emphasized the function of organizing labor, controlling the market, paying wages, and owning the means of production. In Parisian consumer goods industries the manufacturer did indeed organize labor, though with relatively little control over workers because so many worked at home. Additionally, the manufacturer controlled the selling of manufactured goods, but he or she also monitored fashion and other changes in consumer tastes.

This examination of manufacturing procedures and developments in wallpaper making, cabinetmaking, and fan making reveals the viability, even the dynamism of hand production in nineteenth-century France. The general rise in consumer demand, the expansion of a consuming population, and innovations in science and technology clearly did not preclude alternatives to the factory system of production. Manufacturers and workers in these and other industries engaged in a variety of production processes and structures that satisfied consumer demand for tasteful, fashionable, and artistic goods within the framework of hand manufacturing. The evidence presented here should call into question the assumption, implied or explicit, that hand, parcelized, and small-scale manufacturing represented stagnation or decline in French industry. Indeed, many entrepreneurs in consumer goods industries responded creatively and successfully to changes in the market and in consumer tastes by developing the possibilities of hand manufacturing.

It would be wrong to ignore or dismiss other factors in the persistence of hand manufacturing in nineteenth-century France, such as demography (the slow growth of a consuming population), a sizable landed labor force, and entrepreneurial hesitation before the large capital outlays that factory production required. But consumer taste clearly has a place, too, in explaining the persistence of hand manufacturing in nineteenth-century France. In all of the industries studied here, monitoring taste and fashion became an increasingly important function of the manufacturer as the standard of living rose (at least among some portion of the population) and bourgeois consumers could exercise more taste and discrimination in what they purchased for home furnishing and dress.

1. Long view of the exterior of the Crystal Palace. Reprinted from John Tallis, *Tallis's History and Description of the Crystal Palace*, vol. 1 (London and New York: John Tallis and Co., [1852]), facing title page.

2. Exterior of the Crystal Palace, north transept. Reprinted from Tallis, *Tallis's History*, vol. 3, facing title page.

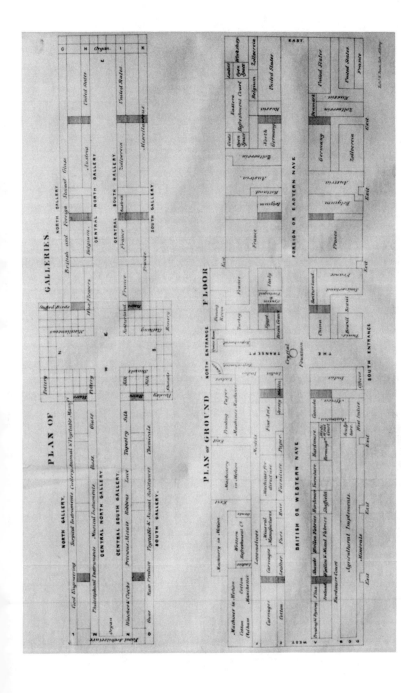

3. Floor plan of the Crystal Palace. Reprinted from Great Exhibition of the Industry of All Nations, 1851, *Report of Benj. P. Johnson, Agent of the State of New York* (Albany, N.Y.: C. van Benthuysen, 1852), facing p. 7.

4. Interior of the Crystal Palace, main avenue looking east. Reprinted from Tallis, *Tallis's History,* vol. 3, p. 51.

5. Silver dressing table by Froment-Meurice. Reprinted from The Art Journal, *The Art Journal Illustrated Catalogue: The Industry of All Nations, 1851* (London: J. Virtue, 1851).

6. Carved walnut sideboard by Fourdinois. Reprinted from The Art Journal,
Illustrated Catalogue.

Stéréoscope pour voir les dames de près.

7. "Stereoscope for viewing women up close." From Amédée Charles Henri Noé, *L'Exposition de Londres croquée par Cham,* 2e promenade (Paris: Martinet, 1851). Boston Public Library, Fine Arts Department; reproduced courtesy of the Trustees of the Boston Public Library.

LA GRANDE FONTAINE DE L'EXPOSITION.

— Je vais pousser ma femme là dedans : cela m'évitera la peine de la conduire aux eaux.

8. "The Great Fountain of the Exhibition." From Noé, *Exposition*. Reproduced courtesy of the Trustees of the Boston Public Library.

Ce que l'on éprouve au sortir de l'Exposition est un impérieux
besoin de s'asseoir n'importe où.

9. "On leaving the exhibition one feels an urgent need to sit down anywhere."
From Noé, *Exposition*. Reproduced courtesy of the Trustees of the Boston Public
Library.

10. Salon, probably during the early Restoration. Courtesy of the Musée Carnavalet, Paris. Photograph by Joseph Chemtob.

11. Salon during the Second Empire: H. Montant, "By the Fireside." Courtesy of the Musée Carnavalet, Paris. Photograph by Joseph Chemtob.

12. Re-creation of a Restoration bedroom, including wooden bed. Courtesy of the Bibliothèque des Arts Décoratifs, Paris, Collection Maciet.

13. Advertisement for iron beds (1833). Courtesy of the Bibliothèque des Arts Décoratifs, Paris, Collection Maciet.

An Iron Bedstead, and Child's Cot, also of iron, contributed by M. Dupont, of Paris, will attract attention from the rich and elaborate designs which they exhibit, especially the former object. This has a kind of frieze in basso-relievo, running round one of the sides and the end, representing a hunting party ; the flat terminating pillars are also similarly ornamented. The whole is of cast-iron, produced from a mould that brings out the figures and details of the design with remarkable sharpness and decision. The frame-work of the cot is very light and elegant, and the introduction of a young angel at its foot, as if keeping watch over the little sleeper, is a pretty idea : the basket and fringe are made of netted wool. We may, perhaps, be allowed to take an objection to the practical convenience of the bed, although we may unequivocally express approbation of its ornamental design.

14. Iron bed by Dupont. Reprinted from The Art Journal, *Illustrated Catalogue*.

15. Stove heater in a bourgeois dining room during the Restoration. *Interior of a Dining Room,* etching by L.-P. Debucourt, 1821, after a painting by Martin Drolling, 1816. Reprinted from Mario Praz, *An Illustrated History of Furnishing, from the Renaissance to the Twentieth Century,* trans. William Weaver (New York: G. Braziller, 1964).

16. Fireplace in the salon of a Second Empire apartment. Courtesy of the Bibliothèque des Arts Décoratifs, Paris, Collection Maciet.

17. Moderator lamp by Laroche. Reprinted from Tallis, *Tallis's History,* vol. 3, p. 33.

18. Bronze mantel clock and candlesticks by Lerolle frères. Courtesy of the Bibliothèque des Arts Décoratifs, Paris, Collection Maciet.

Le papier peint : fabrication « à la planche ».

19. Hand method of wallpaper printing (1889). Courtesy of the Bibliothèque des Arts Décoratifs, Paris, Collection Maciet.

La fabrication du papier peint : machine à imprimer.

20. Machine method of wallpaper making (1889). Courtesy of the Bibliothèque des Arts Décoratifs, Paris, Collection Maciet.

21. Floral wallpaper by Zuber. Courtesy
of the Bibliothèque des Arts Décoratifs,
Paris, Collection Maciet.

22. Stag-hunt wallpaper by Delicourt. Courtesy of the Bibliothèque des Arts
Décoratifs, Paris, Collection Maciet.

THE ROYAL FAN
BY P. DUVELLEROY, PARIS

FRENCH FAN
BY P. DUVELLEROY, PARIS

23. Fans by Duvelleroy. Reprinted from Tallis, *Tallis's History,* vol. 1, p. 218.

Chapter Five

Flexible Specialization in Luxury and Art Industries

Early in the 1840s Charles Christofle (1805–63) started a new industry in France. This young and ambitious craftsman, winner of a gold medal for jewelry at the 1839 industrial exhibition in Paris, seized the opportunity to initiate and monopolize large-scale manufacturing of silver-plated and gilded items. He bought the patents to an electrolytic metal-plating process, and began producing silver-plated table flatware and gold-plated decorative objects in unprecedented quantities. Metal plating itself was not new; ever since antiquity artisans had applied a thin layer of gold or silver to base metals in order to produce less expensive but still fine-looking articles of adornment and household furnishing.[1] But the new electrolytic procedure transformed the French goldsmithing and silversmithing industry (*l'orfèvrerie*) and provided bourgeois consumers with a wide range of beautifully plated items, remarkable for the durability of their rich appearance and for their comparatively low cost. "Women should rejoice," one women's periodical asserted in a puff for the electroplated goods sold by Messieurs Boisseaux and Detot; no longer would reasons of economy force women to buy "coarse [*grossier*] table flatware made of iron." The new industry put at consumers' disposal relatively cheap spoons and forks—75 to 84 francs a dozen, depending on the amount of decoration—that resembled the finest silverware in their elegance, polish, and finish. The "charming" gilded and silver-plated

1. Commission française sur l'Industrie des Nations, *Exposition universelle de 1851: Travaux de la Commission française sur l'Industrie des Nations* (Paris: Imprimerie impériale, 1855) 6:103–24; C. Laboulaye, *Encyclopédie technologique: Dictionnaire des arts et manufactures de l'agriculture, des mines, etc.,* 2d ed. (Paris: Lacroix-Comon [1853?]), vol. 2, sect. 4 of "Orfèvrerie."

products at Boisseaux and Detot, the writer concluded, "will tempt many women."[2]

Gold and silver products—teapots, goblets, cruets, trays, spoons, forks, salt cellars, snuff boxes, brooches, bracelets, necklaces, jewelry caskets, and so forth—had a long history as possessions of royalty, the church, and the aristocracy. But the value of the raw materials and of the skill and talent that went into the manufacture of such items priced them beyond the means of all but the wealthiest of the bourgeoisie. Starting in the eighteenth century, however, new techniques of hand manufacture reduced the amount of labor involved in some aspects of goldsmithing and silversmithing, thus increasing the output and lowering the cost of certain products.[3] In addition, the process of fusing a thin layer of silver to a sheet of copper was perfected and successfully exploited by the Sheffield works in England.[4] Craftsmen applied basically the same manufacturing procedures to this silver-plated copper as they did to pure silver, but the finished products were less expensive. A growing bourgeoisie provided an avid market for these cheaper gold, silver, and silver-plated articles—a fact apparently not lost on the jewelry maker and *orfèvre* Christofle.

But what did a dramatic technological change like electroplating mean for French consumers and producers? Did the success of electroplating imply a decline in bourgeois taste, with elegance and taste sacrificed in favor of low cost? Does the example of *orfèvrerie* contradict the arguments of the preceding chapter, that bourgeois taste for beautiful and stylish furnishings promoted hand methods of manufacturing at the expense of mechanization?

This chapter answers these questions by examining, from a consumer perspective, production methods in goldsmithing and silversmithing, and in another industry that experienced significant technological change under the July Monarchy, bronze making. It suggests that in adopting particular scientific and mechanical means of manufacturing, *orfèvres* and bronze makers were not only increasing efficiency and output but also accommodating consumer demand for fashion, art, and quality in household furnishings. The kinds of

2. *Gazette des femmes,* 1 March 1845, 16.
3. Serge Grandjean, *L'Orfèvrerie du XIXe siècle en Europe* (Paris: Presses universitaires de France, 1962), 38; Laboulaye, *Encyclopédie.*
4. John Culme, *Nineteenth-Century Silver* (London: Hamlyn, 1977), 10–18.

technological changes that occurred in *orfèvrerie* and bronze making in the 1840s more closely approximate Charles Sabel and Jonathan Zeitlin's notion of "flexible specialization" than the classical and Marxist concepts of inevitable standardization resulting from mass production methods.[5] That is, electroplating in goldsmithing and silversmithing, and the Collas method of reproducing statues in bronze making, allowed entrepreneurs to vary their product lines as well as to increase their output and lower their production costs. The success of Christofle and that of the bronze maker Ferdinand Barbedienne (1810–92) rested on their ability to alter quickly the styles and designs of their products to meet the changing tastes of a growing number of consumers.

Orfèvrerie and bronze making differed considerably from capital goods industries and even from most other consumer goods industries because of the high value of the primary materials involved and the importance of art in the manufactured product. Both of these issues were paramount in Duke Albert de Luynes's report on precious-metal products at the Crystal Palace exhibition, and they help explain the development and consumer appeal of the new technologies. Electroplating and the Collas technique generated products that closely resembled solid gold or silver goods, or hand-sculpted works of art, respectively. These superb imitations cost considerably less than pure or original objects, but they were by no means so cheap as to be available to a mass market. Indeed, part of the appeal of such goods for middle-class consumers was the obvious amount of valuable and rich-looking gold, silver, or bronze they contained. Furthermore, despite the vastly increased output made possible by technological innovations in *orfèvrerie* and bronze making, the articles thus produced still had substantial artistic quality. Christofle and Barbedienne won praise and awards at the exhibition because of the artistic

5. Charles F. Sabel and Jonathan Zeitlin, "Historical Alternatives to Mass Production: Politics, Markets and Technology in Nineteenth-Century Industrialization," *Past and Present* 108 (1985): 133–76. See also Michael J. Piore and Charles F. Sabel, *The Second Industrial Divide: Possibilities for Prosperity* (New York: Basic Books, 1984); Alain Cottereau, "The Distinctiveness of Working-Class Cultures in France, 1848–1900," in *Working-Class Formation: Nineteenth-Century Patterns in Western Europe and the United States*, ed. Ira Katznelson and Aristide R. Zolberg (Princeton: Princeton University Press, 1986), 111–54; Alain Faure, "Petit Atelier et modernisme économique: La Production en miettes au XIXe siècle," *Histoire, économie et société* 4 (1986): 531–57.

design of their products and the finish that skilled craftsmen added to them.

This chapter suggests that the nature of the products of *orfèvrerie* and of bronze making, and the meaning of these goods for bourgeois consumers, required that manufacturers uphold high standards of art and taste even when they introduced technologically advanced methods. Indeed artistic merit and tastefulness were even more important to business success when the elements of pure precious metal or handmade uniqueness disappeared. The manufacturers at the Crystal Palace exhibition who were most celebrated were those who best combined entrepreneurial innovation with an appreciation for art and taste, a not impossible but nonetheless problematic combination in France in 1851.

THE ROYAL CRAFTS OF GOLDSMITHING AND SILVERSMITHING

A much-admired example of silversmithing in the Crystal Palace was a silver dressing table by François Désiré Froment-Meurice (1802–55), the son of an *orfèvre,* who first won recognition for producing a tasteful but low-cost tea service and set of plates and later moved on to more artistic, highly decorated, expensive items. Commissioned by an association of women Legitimists for the duchess of Parma, this dressing table represented the height of royal patronage of the skilled crafts, and the opulence of aristocratic material life. Made of silver with extensive inlay work, the dressing table was covered with traditional symbols of royalty—lilies, pennants with fleurs-de-lis, crowns, crosses, shields, angels, cherubs, and medieval knights and saints. Accompanying the dressing table were candle holders, a cruet and salver, and jewelry caskets, and the whole was backed with a mirror some two to three feet high and adorned with sculpted figures and designs (fig. 5).

This extraordinary piece of silversmithing required six years and many talented hands for its manufacture. Froment-Meurice hired a panoply of artists and craftsmen, including an architect to design it, two well-known artists to model and sculpt the thirty-one figures decorating the set, a woodcarver to form the floral ornamentation, and other artisans to provide the inlaid enamel work.[6] The practice of

6. The Art Journal, *The Art Journal Illustrated Catalogue: The Industry of All Nations, 1851* (London: J. Virtue, 1851), reprinted as *The Crystal Palace Exhibition*

commissioning several architects and artists to contribute to a single work of *orfèvrerie* dated at least from the First Empire, when classical style and monumental effect were fashionable. Architectural styles influenced designs in *orfèvrerie,* so it was common for *orfèvres* to work with architects in the creation of decorative gold and silver pieces. With the Restoration, however, the new romantic emphasis on nature and on medieval styles led to a shift in *orfèvrerie;* instead of architects, artists and sculptors were hired to ornament large works.[7] Long gone were the days of Renaissance craftsmen like Benvenuto Cellini who combined artistic creativity with technical skill to produce, singlehanded, masterpieces of art in gold and silver. Though plenty of royal and aristocratic patrons remained for *orfèvres* and their associates to serve, the new markets for the products of *orfèvrerie* that burgeoned in the eighteenth and especially nineteenth centuries required changes in manufacturing procedures.[8] Even the fabulous Froment-Meurice dressing table represented a model for less exalted consumers, according to Émile Berès's report for *L'Illustration.* Berès suggested that any young girl could be a "princess" by obtaining a dressing table with artistic decoration resembling that of the Froment-Meurice production, though undoubtedly it would be made of less expensive material than silver. The true value, Berès indicated, was in the art and style of the piece, which could be reproduced in other materials and at lower cost while continuing to satisfy consumers' desire to appear tasteful and cultivated, and so, in a sense, noble.[9]

Historically, *orfèvrerie* consisted almost entirely of alternately beating and heating gold or silver into the desired shape. According to illustrations in a 1788 technical encyclopedia, *orfèvres* habitually worked at specially designed work benches upon which were mounted small, rotating wooden mandrels or anvils. Leather pouches hung beneath the mandrels to catch bits of precious metal as they were hammered or cut away; these pieces could later be reused. Beating the metal on the mandrel required great skill and dexterity to maintain a consistent thickness of the gold or silver. The next basic step in the manufacture of a vase or salver was decorating the piece. The most common means of decoration—and the most artistic aspect

---

Illustrated Catalogue (New York: Dover, 1970), 130–31; Henry Bouilhet, *L'Orfèvrerie française aux XVIIIe et XIXe siècles* (Paris: H. Laurens, 1910), 274.

7. Bouilhet, *Orfèvrerie française,* 185–86, 232–36; Grandjean, *Orfèvrerie,* 69.

8. Laboulaye, *Encyclopédie.*

9. *L'Illustration,* 12 July 1851, 27.

of *orfèvrerie*—was the engraving of designs or patterns onto the metal object. Joining pieces together was an important task in the production of three-dimensional objects like cruets or vases; craftsmen sought a perfect and invisible joint through the correct application of precious and base metals, borax, and heat. The finishing procedures consisted of burnishing the item to a dull glow (an appearance that French consumers favored), or polishing it to a high shine (a preference among the English). The basic tools for this method of manufacturing were hammers, anvils, files, chisels, compasses, a blow lamp, and a forge.[10]

In the late eighteenth century growing demand for goldsmiths' and silversmiths' products led to the adoption of a variety of new hand tools to increase the efficiency of manufacturing. Rolling mills to flatten and elongate metal, lathes on which to shape and polish pieces, and engraved dies to cut out and decorate objects all appeared in goldsmithing and silversmithing workshops. A cutting wheel, for example, could be rolled along a piece of metal and hammered down, accurately reproducing the same design many times over, instead of engraving individually each rosette or palm leaf. Popular style and consumer taste influenced the extent to which new tools and machines were used; for instance, implements to reproduce symmetrical designs were more successful with imperial than romantic style. Moreover, English consumers' preference for larger, less delicate, and less finely decorated objects meant that machine techniques were adopted in England sooner than in France. Nonetheless, new technology essentially created an important subspecialty of *orfèvrerie* in France—the manufacture of table flatware, performed by *cuilleristes,* or spoon makers.[11]

Even at the end of the eighteenth century, spoon makers produced spoons and forks entirely by hand, stamping them into shape with a swage. Early in the nineteenth century, however, a three-stage process using hand-powered tools increased the output of table flatware to meet rising demand. First the flattened metal was cut, then it was "prepared"—forged with a hammer into the general shape of a fork

10. Grandjean, *Orfèvrerie,* 38–40; *Encyclopédie méthodique par ordre de matières: Arts et métiers mécaniques,* vol. 4, *Planches du dictionnaire des arts et métiers* (Paris: Panckouke, 1788).

11. Commission française, *Exposition* 6:39–41; Grandjean, *Orfèvrerie,* 69; Bouilhet, *Orfèvrerie française,* 186; Laboulaye, *Encyclopédie.*

or spoon. The third stage was inserting the prepared metal into a fly press, similar to devices used in making coins, to stamp out a nearly finished product. A worker then filed away excess silver, and cut out the teeth for forks. This highly divided method of hand manufacture changed drastically in 1840 with the adoption of a machine that other metal workers had been using for several years. This hand-powered, and eventually steam-powered, machine combined the procedures of flattening or rolling the metal and stamping out the form of fork or spoon with an ornamental design. By using well-cut dies made of steel and patterns designed by famous *orfèvres* or suited to a rich customer's desires, these machines could produce beautiful, perfectly matching sets of table flatware at much lower cost and with far fewer workers than before. The value of these machines increased even more in the late 1840s with the use of electroplated nickel silver, which began to replace both the Sheffield sheets of copper/silver and the solid silver used for table flatware in restaurants and homes.[12]

FRENCH *ORFÈVRERIE* AT THE CRYSTAL PALACE EXHIBITION

At a much slower pace machine technology was altering the manufacture of large gold and silver items—vases, cruets, serving platters, teapots, coffeepots, centerpieces, and so forth.[13] Such items were the traditional products of goldsmithing and silversmithing, and they were among the greatest attractions at the Crystal Palace exhibition, both for ordinary spectators and for the press. Analyzing the quality of *orfèvrerie* products and developments in their manufacture was of paramount interest to French industrial experts, who wished to gauge the performance of French industry compared to that of other

12. Commission française, *Exposition* 6:44–45; Laboulaye, *Encyclopédie*. According to the Paris Chamber of Commerce's survey of industry, eighteen spoon makers operated in Paris in 1847, employing 280 workers; the number of workers declined to 111 in 1848. By Parisian standards these were fairly large operations, counting an average of fifteen workers per establishment during times of relative economic prosperity. Commission française, *Exposition* 6:46–47.

13. Jewelry making was yet another branch of *orfèvrerie* where stamps and dies were used. In France imitation jewelry for bourgeois customers was of fairly good quality; designs were copied from the finest original jewelry, and the imitation differed from the original only in the use of hand tools and machines. Laboulaye, *Encyclopédie*.

countries and to determine which trends were advantageous to economic growth and industrial competitiveness in France.

The French jury member for *orfèvrerie,* Duke Albert de Luynes,
was hardly an unbiased judge of these French products. An aristocrat
and prominent patron of the arts, Luynes was obviously apprehensive
about new techniques in *orfèvrerie* and their, in his view, nefarious
effects on style and quality. Changes in the organization of work—
that is, from the *orfèvre* as artist and producer for specific clients to the
orfèvre as manager of an enterprise—were even more disturbing to
Luynes, for he considered the new practices inimical to art. Nonetheless, precisely because Luynes was a connoisseur of gold and silver
objects (among many other types of arts and crafts), his perspective
was that of the wealthy consumer rather than the entrepreneur or the
artisan. Luynes's report on *orfèvrerie* analyzed all of the major French
exhibitors and their contributions to the craft; while acknowledging
the importance of technological innovation and entrepreneurial acu-
men to success in *orfèvrerie* he reserved his highest praise for the
craftsmen who upheld or retrieved traditional skills and artistic taste.

Prominent among French *orfèvres* at the exhibition was Charles
Odiot, son of Claude Odiot, *orfèvre* to the first Emperor Napoleon.
When Charles was young his father sent him to England to work as
a sculptor and to study goldsmithing and silversmithing. Charles returned to Paris with imported styles, tools, and techniques which he
implemented when he took over his father's shop in 1827. Among the
English tools Odiot introduced in his workshop were a round and an
oval lathe, machine tools, cutters (shears), a fly press, and die stamps.
The technique which the fly press and dies made possible—that of
stamping parts of a piece and soldering them together—reduced
production costs but required styles that were more ponderous than
the light and delicate designs that characterized late-eighteenth-
century gold and silver articles.[14]

This modification clearly informed Charles Odiot's personal views
on *orfèvrerie;* he held that lavishness, utility, and precision, rather than
refinement, beauty, or grace, were paramount in the trade. He declared that *orfèvrerie* should be "an industry of opulent usage and
distinguished by perfect execution, by accommodating and useful
forms, by the intrinsic value of the material used with skill."[15] Luynes

14. Bouilhet, *Orfèvrerie française,* 158–61.
15. Commission française, *Exposition* 6:62.

duly praised the young Odiot's well-equipped workshop and his rationally directed, perfectly manufactured work; but he was somewhat unsympathetic to Odiot's pragmatic and efficient approach to his craft. Moreover, Luynes found that Odiot was adulterating or even replacing good French taste with the less refined English preference for solidity and bulk.[16] Alfred Busquet, writing for *La Semaine,* distinguished between Odiot's success as an entrepreneur and his weakness as an artist: "M. Odiot concerns himself excessively with his [business] interests. He will undoubtedly sell his tableware service, while his colleagues exhibited without selling a thing, but he will receive neither the praise nor the sympathies of the artist."[17]

Other artisan manufacturers upheld more traditional methods and styles in *orfèvrerie,* which were noted and admired at the exhibition. According to Luynes, the Prussian immigrant Charles Wagner was more successful than young Odiot at changing and improving style in *orfèvrerie* because he took his inspiration from the past. Wagner revived the difficult art of repoussage—the technique of pressing or hammering a design from the reverse side of a piece. He also demonstrated great skill and innovation in his use of enamel inlay and in the design, modeling, and chiseling of his works in gold and silver. Luynes reserved his greatest admiration, however, for the *orfèvre* Antoine Vechte (1800–68), remarkable for being the only one in Europe at that time who designed and executed his works entirely alone, as the Italian masters of the Renaissance had done.

Vechte's history was representative of the ideal rise of the artisan from obscurity to renown solely on the basis of skill and imagination (like the potter Avisseau, discussed in chapter 1). Orphaned at the age of eleven, Vechte left his natal town of Avallon in the Yonne department to find work in Paris, as an apprentice to an engraver and then to a metal caster. He essentially taught himself modeling; making models for a bronze maker was, for a while, his chief source of support for his family of eleven children. Shortly after beginning his fruitful association with the artist Jean Feuchères (1807–52), Vechte produced articles of bronze and gold—buckles, helmets, silver plates, and small vases—in imitation of Renaissance styles; he sold them

16. See the quotation in chapter 1 from *Dickinson's Comprehensive Pictures of the Great Exhibition of 1851* (London: Dickinson Bros., 1854).

17. *La Semaine,* 31 May 1851.

anonymously.[18] Dealers noticed the exquisitely repoussé objects with sixteenth-century characteristics as well as a distinctive quality, and they finally persuaded Vechte to exhibit under his name in art salons of the 1840s. By the time of the 1851 exhibition Vechte was a very successful artisan manufacturer, and the daughter he trained was also well known for her sculpting and chiseling of precious metals.[19]

Luynes noted Vechte's important innovations in repoussage, which both reduced labor and improved quality. Formerly, goldsmiths roughed out and completed difficult pieces in plaster before producing the whole in gold. Vechte, however, stamped out the separate pieces in metal troughs, then he fastened, soldered, and chiseled them; and he used a notably high quality of solder. Vechte's method of producing the pieces separately required less hammering, so the metal had a more even thickness and there was less danger of holes and tears.[20]

As long as artisans used hand methods and traditional standards, the quality of their products was relatively easy to assess. Rarely could inferior materials and shoddy workmanship pass for something better. Yet experiments in electrolysis generated an entirely new goldsmithing and silversmithing industry almost overnight, in which entrepreneurship was at least as decisive for success as artistic imagination and manual skill, and in which measures of worth were more murky and debatable. A figure who best represented this transition between artisan and entrepreneur was Charles Christofle.

THE TRANSFORMATION OF ORFÈVRERIE

Christofle came from a Lyon silk-manufacturing family which was ruined after the invasion of 1814. The boy had to interrupt his high school studies and learn a trade, so he entered his brother-in-law's jewelry shop as an apprentice. After three years he became a worker, and one year later he joined the management of the Calmette establishment. In 1831, at the age of twenty-four, Christofle headed the

18. Whether Vechte was deliberately trying to deceive dealers and consumers by passing off his productions as antiques is not clear. Luynes suggests no intent to defraud; but false claims to antique origins for such items were common, as they were in the furniture trade.

19. Bouilhet, *Orfèvrerie française*, 204–6; Commission française, *Exposition* 6:74–75.

20. Commission française, *Exposition* 6:74–77.

largest jewelry manufacture in France, and he won a gold medal at the 1839 exhibition for his jewelry, gold and silver filigreed ornaments, and gem works. Early in the 1840s he borrowed money from friends to purchase the Elkington and Ruolz patents on electrolytic gold and silver plating. He explained his starting of this path-breaking endeavor by saying that "it belonged to a man who owed his fortune to industry to apply part [of that fortune] to the execution of this beautiful discovery."[21]

This "beautiful discovery" represented a significant improvement over previous methods of metal plating. In the past, covering items with a thin layer of gold required the use of mercury, which endangered the health of workers. Silver plating was less dangerous, basically entailing the placement of silver leaf on a heated metal surface, followed by burnishing to reinforce the natural solder. The best-known silver-plating process was that developed at the Sheffield works in England (described earlier in this chapter). None of these hand procedures, however, could guarantee a completely even metal plating, and with heavy use the silver often rubbed off to expose the copper beneath. Experiments with electroplating flourished in the 1830s, but they failed to produce beautiful and permanent silver and gold plating. Indeed several *orfèvres*, citing the need to meet consumer demand for variety and to end the importation of silver plate from England, sent written requests to the minister of agriculture and commerce for loans to develop new machinery and techniques for silver and gold plating.[22] The breakthrough was Elkington and Ruolz's combining of gold and silver salts with potassium cyanide, or, for gilding, with gold sulfide and potassium sulfide. This innovation made electrolytic metal plating successful.[23]

In Christofle's Paris shops twelve huge 650- to 700-liter vats held a solution of silver and potassium cyanide. Large strips of fine silver were suspended in the solution at one end of a vat and connected to the positive pole of an Archerean battery. At the opposite end silver

21. Quoted in Bouilhet, *Orfèvrerie française,* 243. The English cousins Henry and George Richards Elkington publicized the electrolytic plating procedure around 1836; it was improved in 1840 by Baron Henri de Ruolz-Montchal. Culme, *Nineteenth-Century Silver,* 116.
22. Archives Nationales, Ministère de l'Agriculture et du Commerce, F¹² 2265, Orfèvrerie, bijouterie, bronzes, fabriques de plaqué d'or et d'argent, an III–1848, 1807–21.
23. Grandjean, *Orfèvrerie,* 41; Commission française, *Exposition* 6:103–9.

strips were placed horizontally, parallel with metal rods connected to the battery's negative pole. From these metal rods were suspended the brass or copper items to be silver plated, which had been thoroughly scoured in alkaline and acid lyes and passed through distilled water before immersion in the vats. Each vat deposited 1,500 to 1,800 grams of silver in twenty-four hours, and it was possible to plate six dozen table settings at a time. The plated pieces were removed from the vats, washed, dried, weighed to confirm the silver weight, and then taken to other shops for burnishing. Gold plating or gilding procedures were similar, the main difference being that the solution was practically boiling, and that there were two different qualities of gold baths. Christofle also used steam-powered lathes and other modern equipment in his shop, which employed some three hundred workers.[24]

Not all of Christofle's colleagues shared his enthusiasm for the new plating process. For one thing, workers and manufacturers who plated metal in the old manner lost jobs and business with the success of the electroplating method.[25] Moreover, as the patent holder, Christofle had the exclusive right to use the new process, thus preventing any direct competition within France at the time of the exhibition. The new technique of metal plating, Christofle's monopoly over it, and his other business practices all raised serious issues at the Crystal Palace exhibition regarding technological development and the artistic quality of manufactured goods in France. Were mechanical and scientific innovations in the production of artistic and/or luxury goods compatible with high standards of beauty and quality in the manufactured products? And what was the best means of guaranteeing continued high standards—government regulation, or the honor of a master craftsman?

French industrial experts commenting on the exhibition were generally optimistic that new technologies could and in fact did enhance the quality of gold, silver, and plated products. As Luynes noted in his official account of *orfèvrerie* at the exhibition: "Today, with a very elevated sense of the artistic, our *orfèvres* have more resources, and

24. Commission française, *Exposition* 6:120–24.
25. Ibid., 126–27. Five years after the exhibition, Audiganne claimed that gilders and silver platers who used the old methods were wiped out. Armand Audiganne, *L'Industrie contemporaine, ses caractères et ses progrès chez les différents peuples du monde* (Paris: Capelle, 1856), 51.

[they] easily overcome the difficulties that cost so much effort and time in earlier periods."[26] Similarly Armand Audiganne, writing on the occasion of the 1855 exhibition in Paris, maintained that new methods of manufacturing, including electroplating, were in fact contributing to higher quality in the products of *orfèvrerie*. For example, Audiganne explained how, due to improved soldering techniques, individual craftsmen could produce beautiful repoussé work by working on ten to twelve separate pieces and then joining them into a nearly finished product. "From the point of view of execution, the new system attains as great perfection as the old, [but] it is simpler and much quicker."[27] According to Audiganne, the division of labor among designers, sculptors, engravers, and many other workers was not at all harmful to the craft or to the quality of its products; in fact he considered that the artistic ability and technical skills of *orfèvres* were improving.[28]

Luynes, by contrast, was much more ambivalent about the effects of divided labor; he discerned a transformation in the role of the master *orfèvre* from craftsman to manager of labor as the craft became more commercialized.

> Having become manufacturers [*industriels*], [goldsmiths and silversmiths] no longer believe that they themselves must be producers and artists. Division of labor, so useful for manual labor, has overtaken intellectual work. The capable *orfèvre* of today receives a commission, conceives of the whole, then, running to the imagination and labor of others, he gets workers and artists of the city to compose, design, mold, forge, and engrave; his role in the job, as entrepreneur, consists of the preparation of the different pieces of the product, the adjusting, the assembling by soldering, the burnishing or polishing of the united parts, and the final mounting. He is simply an entrepreneur of works of art, intelligent, tactful, and tasteful.[29]

Luynes, quoting from the Paris Chamber of Commerce's survey of industry of 1848, was clearly concerned about the potential decline of the art and craft when master *orfèvres* were themselves so detached from the manufacturing process. Yet he could still point to extremely

---

26. Commission française, *Exposition* 6:39.
27. Audiganne, *Industrie contemporaine,* 38.
28. Ibid.
29. Commission française, *Exposition* 6:53.

successful entrepreneurs who produced beautiful works of art even with divided labor. The key to good *orfèvrerie* in the middle of the nineteenth century, according to Luynes and others, was the training and ability of the individual craftsman/entrepreneur. "Almost all [goldsmith and silversmith entrepreneurs] have worked with their own hands and give to their businesses the practical knowledge they acquired as workers or as masters."[30] Echoing this sentiment, the reporter for *L'Illustration* attributed Christofle's success as much to his craft heritage as to his clever exploitation of electroplating: "Far from restricting himself to the able application of the patented procedures, M. Christofle remembers the early studies of his first industry [jewelry making] . . . and he is committed to imprinting with the cachet of his practical experience in *orfèvrerie* and jewelry making the base metals destined to receive . . . the layer of gold or silver."[31]

Christofle not only upheld the honor of *orfèvrerie* by producing well-designed, beautiful pieces of gold, silver, and plated metal; he also measured precisely the amount of gold or silver on his plated goods. Before and after the immersion of table flatware or other items into the plating vats, Christofle carefully weighed the items, and he thus was able to calculate the amount of silver or gold that was transferred to each. Earlier methods of metal plating could not guarantee such accurate accounting; and Luynes characterized this practice of Christofle's as an indication of his honor and trustworthiness as a craftsman. Indeed, Luynes feared that government intervention in electroplating—by repurchasing the patent and issuing production standards—would destroy the integrity of *orfèvres* like Christofle and eliminate public trust in the quality and precious-metal content of plated goods.[32]

Despite technological innovation and commercialization, French industrial experts in the middle of the nineteenth century by and large agreed that individual craftsmen at the head of *orfèvrerie* enterprises were primarily responsible for the success of the industry in a competitive setting like the Crystal Palace exhibition. Increased public demand for table flatware, serving dishes, jewelry, and decorative boxes propelled goldsmiths and silversmiths to increase output and adopt new methods of production, but not necessarily at the expense

30. Ibid.
31. *L'Illustration*, 14 June 1851, 380.
32. Commission française, *Exposition* 6:118.

of art, taste, and quality. In accommodating a range of consumers increasingly disparate in their disposable income and artistic taste, *orfèvres* expanded their range of products from unique, totally hand-made, and richly engraved items of solid silver or gold, to easily reproducible, less decorated, plated pieces manufactured with the aid of hand tools, machines, and electrolytic procedures. At all levels consumers demanded variety, beauty, fashion, and elegance; techno-logical innovations had to respond to changes in style and to the demand for rich appearance.[33] The case of *orfèvrerie* further illustrates the need for a more nuanced understanding of the relationship be-tween demand and technological change in manufacturing, an under-standing that accounts for flexible specialization rather than assuming a universal trend toward standardization and mass production. An-other luxury industry whose nineteenth-century trajectory typifies this kind of development was the manufacture of bronze statues.

ART FOR THE HOME: BRONZE STATUES

A major figure among a large and varied cast of characters in Balzac's *La Cousine Bette* is the artist Wenceslas Steinbock. In this novel of greed, lust, and betrayal Wenceslas initially appears to be a fortunate recipient of the spinster Bette's possessive and self-interested charity. Poor, unemployed, and demoralized, the beautiful Polish immigrant attracts the interest of his neighbor Bette, an industrious needle-woman who supplements her meager income by sponging off her affluent relatives, the Hulot family. Bette rescues Wenceslas from suicide and launches him on the road to earning a living. She loans him money for living expenses, puts him in contact with a bronze maker and dealer, and commands him to work regularly and for long hours. Over a period of five years Wenceslas manages to produce a bronze statue, a silver seal, and a model for a clock; but he fails to earn any money, despite Bette's tyranny. Bette's reward for her questionable investment is a passive and grateful young craftsman who knows that he owes his very life to the energetic, domineering, and duplicitous woman. But professional success changes this

33. Ibid., 44; Jeanne Gaillard, *Paris, la ville, 1852–1870* (Paris: H. Champion, 1977), 438–46. According to Laboulaye, *Encyclopédie,* progress in *orfèvrerie* must necessarily combine lower costs with a higher proportion of value in art work.

relationship, and it changes the characters of Bette and Wenceslas; or rather, the two revert to type.

Bette's beautiful niece Hortense Hulot discovers the bronze sculpture by Wenceslas in an antique dealer's shop. A young woman of good taste and artistic sensibility, Hortense considers the statue a fine piece of art and buys it with her savings. Upon finally meeting, Wenceslas, Hortense falls in love with the artist. The two young people marry, and Wenceslas's career surges upward, with several commissions for public monuments and private works from government ministers associated with the well-connected Baron Hulot, father of Hortense. Having lost her protégé, Bette determines to ruin the Hulot family, and she nearly succeeds. Wenceslas, deprived of Bette's tyrannical discipline and softened by the unquestioning love of an adoring wife, betrays his great promise as an artist. His statue of a famous marshal of France receives public criticism, and he sinks into a life of idleness, incapable of working on even the slightest bronze trinket.[34]

Balzac, of course, was more interested in explaining the sad decline of a once-proud family than in examining the fine points of bronze making, but he does offer some insights into the life of a sculptor and the production of industrial art under the July Monarchy. Wenceslas has true artistic genius, and he rightfully aspires to the lofty profession of sculpting large statues for public display. Such artists were richly rewarded by the state and enjoyed high social status. As part of his commission for the government to make a statue of the marshal, Wenceslas is given a studio and housing in addition to payment for the sculpture, and he gains entrée into the most prestigious salons. But in reality few artists were so fortunate as the fictional Pole, and government officials withdrew their favors as suddenly as they granted them. More commonly, artists and craftsmen spent their lives producing the decorative objects for the home that Wenceslas made at the beginning of his career.

Some scope for artistic creativity remained in the production of statuettes, grouped figures, strongboxes, vases, stemmed cups, bowls, and other objects made of bronze. But in his report on bronze making for the 1827 Paris exhibition of industry the economist Adolphe Blanqui criticized bronze manufacturers for emphasizing

---

34. Honoré de Balzac, *La Cousine Bette,* vol. 17 of *Oeuvres complètes de M. de Balzac* (Paris: Les Bibliophiles de l'originale, 1968).

works of art that only a tiny, wealthy elite could afford to buy. "Work for the people who buy every day," Blanqui admonished bronze makers (along with cabinetmakers), "rather than for those who buy only once."[35] Perhaps Blanqui wished to see bronze makers expand their product line to more mundane household furnishings that every bourgeois family needed for basic comfort and well-being: mantel clocks and candle holders (see fig. 18), consoles, shelving, hanging lamps, sconces, table lamps, centerpieces, andirons, fenders, paper knives, ink stands, and so forth. Bronze makers apparently heeded Blanqui's call for more affordable products—or they simply responded to increased demand. As early as 1839 a reporter for the industrial exhibition of that year targeted bronze making as a growth industry: "It becomes clear that [bronzes] tend to penetrate into middling incomes and to become objects of ordinary consumption; there lies the secret of [bronze making's] fortune and the guarantee of its future."[36]

Bronze making, like goldsmithing and silversmithing, consisted of several operations performed by different workers: designing, modeling, melting and pouring, lathing, engraving, assembling and mounting, sculpting, and gilding. These operations were sometimes concentrated in one workshop, or they were dispersed among independent and specialized workers in a neighborhood. The manufacturer himself was responsible for coming up with a model of the object to be cast in bronze, which an artist executed out of wax, plaster, or wood. This model was then used to create a mold (in separate pieces for complex objects) in special sand. After a worker removed the model, the sand was heated and dried in an oven to receive the molten bronze. The correct composition of the bronze alloy was essential to the success of the operation: 75 percent red copper, 22 percent zinc, 2 percent tin, and 1 percent lead. Once the poured bronze had hardened, a chiseler (*ciseleur*) worked on the pieces to eliminate imperfections from the molding process and restore the appearance of the original model. The chiseler's task required artistic

---

35. Adolphe Blanqui, *Histoire de l'Exposition des produits de l'industrie française en 1827* (Paris: Renard, 1827), 155.

36. Exposition des produits de l'industrie française en 1839, *Rapport du jury central* (Paris: Bouchard-Huzard, 1839) 3:20–21. See also Charles Laboulaye, *Essai sur l'art industriel* (Paris: Bureau du dictionnaire des arts et manufactures, 1856), 144.

appreciation for the work as a whole, in addition to manual skill; so much so that the bronze makers' association founded a school of design in 1837 to ensure artistic quality in bronze. The next phase of bronze making was the assembling and mounting of the pieces, including the filing of joints to disguise all impurities and signs of divided manufacture. A gilder then washed the whole object in acid solution, brushed it, and applied gold, mercury, and heat to obtain a shiny, permanent gold surface. He, or a colorer, could then burnish the object to a dull glow, or tint the gold to different shades of color through the use of other acidic solutions.[37]

TECHNOLOGICAL CHANGES IN BRONZE MAKING

Major developments in bronze making in the eighteenth century contributed to the expansion of the industry by producing more objects for private consumption and home furnishing. The gilding technique described above, for instance, made bronzes more desirable for home decoration, and a procedure for reproducing bas-reliefs on a smaller scale also broadened the market for bronzes.[38] Moreover, in the nineteenth century the lost-wax process of bronze casting, which improved the quality of bronze statues, was revived.[39] However, far surpassing these and other innovations in its contribution to the increase in consumption of bronze items was the Collas technique of reproducing and reducing large statues. In 1836 the mechanic and inventor Achille Collas (1795–1859) joined forces with the wallpaper maker and bronze maker Ferdinand Barbedienne to create a panto-graph device that traced the outline of a statue (the model) and simultaneously passed that design, reduced in scale, onto a plaster block. At the 1839 exhibition Collas and Barbedienne won national acclaim for their small-scale reproduction of the Venus de Milo from the Louvre.[40]

37. Guillaumin, ed., *Dictionnaire du commerce et des marchandises* (Paris: Guil-laumin, 1839) 1:377–78; Audiganne, *Industrie contemporaine,* 75–76.

38. Laboulaye, *Essai,* 137.

39. Ibid., 113, Jeremy Cooper, *Nineteenth-Century Romantic Bronzes: French, English and American Bronzes, 1830–1915* (Newton Abbot, England: David and Charles, 1975), 25.

40. Audiganne, *Industrie contemporaine,* 79–80; Guillaumin, *Dictionnaire* 2:377; Yves Devaux, *L'Univers des bronzes et des fontes ornementales: Chefs-d'oeuvres et curiosités, 1850–1920* (Paris; Pygmalion, 1978), 259.

The man primarily responsible for this transformation of bronze making was Barbedienne. Born in 1810 in the Calvados department, Barbedienne went to Paris at the age of thirteen where he became an apprentice saddle maker. Moving on to work for several different wallpaper makers, Barbedienne started his own wallpaper establishment in 1833; and he met Collas, who had invented a press for printing wallpaper. The two started producing bronzes, using Collas's pantograph technique, in 1838. The business grew, and when Barbedienne became sole owner in 1859 he employed some three hundred workers.[41] The only other bronze maker who approached Barbedienne in influence and output was Denière, who started in the 1820s and who concentrated on objects of furnishing—such as hanging lamps, candelabras, and mantel clocks—rather than works of art.[42] Denière, like Barbedienne, won several prize medals at successive national and international exhibitions of industry, and his shop windows filled with bronze furnishings attracted hordes of customers. But Denière was not really a competitor of Barbedienne because of his explicitly commercial, as opposed to artistic, emphasis.

In a sense the Collas method inaugurated an entirely new industry: the reproduction of famous statues into bronze statuettes for home decoration. The pantograph, with its capacity for transferring the outline of a whole statue onto a plaster block, meant that fewer stages were necessary in the reproduction process; in particular the copying by a sculptor was eliminated. The new technique greatly lowered the price of bronze copies, rendering them affordable to more consumers. And such reproductions were more accurate imitations than those made by sculptors. Armand Audiganne maintained that Collas copies were in fact superior to any imitation by human hands. According to Audiganne, a great artist who copied a famous statue must inevitably inject his own originality into the piece, and a lesser artist might fail to recreate the original sculptor's spirit. For Audiganne, then, the mechanical reproduction was the best means of capturing the form and genius of a great work of art, and of disseminating art appreciation to the masses through the sale of copies.[43]

41. Pierre Kjellberg, *Les Bronzes du XIXe siècle: Dictionnaire des sculpteurs* (Paris: Les Éditions de l'amateur, 1987), 653; Devaux, *Univers,* 259.

42. Kjellberg, *Bronzes,* 659; Audiganne, *Industrie contemporaine,* 82–83.

43. Audiganne, *Industrie contemporaine,* 78–80. See also Laboulaye, *Essai,* 112–13.

Acquiring art was an important symbol of high social status for French bourgeois. It meant that a bourgeois family had satisfied its basic material needs and could afford to engage in the "leisured" activity of art patronage. Since few middle-class men and women had the time or money for serious art collecting, most settled for reproductions, and these became a standard feature of bourgeois households, especially with the development of new and improved reproduction methods. Bronze making was a stunning example of how the production, or rather reproduction, of art became a flourishing industry. "If there is one of our industries which, in terms of perfection . . . enjoys great favor, it is undoubtedly that of bronzes, . . . which appear . . . in all our salons from the most sumptuous to the most modest," the feminine press proclaimed in 1843.[44] A decade later Audiganne confirmed the growing popularity of bronze reproductions, and particularly of reduced copies of antique statues; such a statuette might be the only object of art possessed by a middle-class family. "[B]ronzes of more or less rich type, of more or less irreproachable taste, have already become part of the furnishing of fairly well-off families."[45]

The same questions applied to bronze making as to *orfèvrerie:* were technological changes in manufacturing detrimental to the artistic quality of the products? Did the expansion of the market for artistic goods entail a necessary decline in taste to accommodate less cultivated consumers? The answers were conflicting. Audiganne thought Collas reproduction enhanced artistic quality in bronzes; Luynes thought that the electroplating of bronze statues resulted in gilding that was far inferior to that produced by the earlier method.[46] Both men, as well as other commentators on French industry, evaluated bronze makers individually in terms of their contributions to art and taste in the industry. Audiganne deemed Barbedienne a true craftsman and artist who was capable of defying the popular taste for the ephemeral in his commitment to reproducing only great works of art: "At Barbedienne's establishment the merchant effaces himself, when the need arises, behind the man of taste."[47] According to Audiganne,

44. *Le Petit Messager des modes,* 1 March 1843, 37.
45. Audiganne, *Industrie contemporaine,* 87. Audiganne wrote: "Often in our homes [bronzes] are the only representatives of art." Ibid.
46. Commission française, *Exposition* 6:232.
47. Audiganne, *Industrie contemporaine,* 81.

Barbedienne maintained high standards of quality in all of his productions, he used only the finest raw materials, and he manufactured furnishings appropriate to accompany statues in the home (rather than making only statues, which might not go with a homeowner's decor).[48] Ultimately, experts claimed, the artistic quality of bronze objects was the responsibility of the manufacturer, for he hired and commissioned the different workers, and he alone could judge the quality and tastefulness of a bronze item at the different stages of its production. Both Luynes and Émile Berès, writing for *L'Illustration,* criticized some bronze makers for sacrificing good taste to higher profits.[49] Luynes, for one, detested Louis XV style and the bronze makers who pandered to the popular enthusiasm for this fashion, but he happily noted that tastefulness returned to bronzes around 1849.[50]

Bronze makers themselves, in letters to the Ministry of Agriculture and Commerce under the July Monarchy, manifested concern for the tastefulness of their products as well as the viability of their trade. One petitioner requested that the government provide models of art work for bronze makers, to ensure the artistic quality of their goods. In return he promised to support the government by contributing to a new Louis-Philippe style that would combat the ill effects of the poor examples of Louis XV style inundating Paris.[51] Although this particular petitioner was unsuccessful, the bronze-making industry of Paris received substantial financial aid from the Second Republic in 1848. Bronze making suffered terribly during the depression and the revolutionary years of 1847–48, and like the cabinetmakers of Saint-Antoine, bronze makers received loans from the government to enable them to continue manufacturing during this period. Bronze making employed 416 workers and did 1,920,900 francs' worth of business in 1847; those figures declined to 92 workers and 364,971 francs in 1848.[52]

Bronze making was a unique industry in France in the middle of the nineteenth century in that it existed only in Paris. As many commentators noted, bronze making was a quintessential Parisian

48. Ibid., 81–82.
49. *L'Illustration,* 19 July 1851, 39–40.
50. Commission française, *Exposition* 6:232–33.
51. Archives Nationales, Ministère de l'Agriculture et du Commerce, F[12] 2282, Bronzes, ébenisterie, bijouterie, an IV–1848.
52. Commission française, *Exposition* 6:233.

industry, a product of the art, taste, and talent of a highly skilled urban labor force, surrounded by some of the finest art and architecture in the Western world. France imported no bronzes from abroad; and exports constituted some two-thirds of annual bronze production, despite a 33 percent tariff that other countries charged on French bronzes.[53] Clearly bronze making was essential to the establishment of France's market niche in art and luxury industries.

MANUFACTURERS AND MEN OF TASTE

Audiganne, Luynes, Berès, and other writers on industrial matters in the middle of the nineteenth century suggested that technological developments and expanding markets caused a split in the function of manufacturers. Industrial producers, particularly of luxury or artistic goods, confronted a choice between being entrepreneurs or managers, and being artists or craftsmen. Audiganne wrote regarding artistic bronzes for the home, "It is especially dangerous here [for the manufacturer] to be dominated by a purely mercantile spirit. On the other hand . . . one cannot require of a manufacturer that he rigorously devote himself to art for art's sake. Between these two extremes is a middle way that a man of taste knows how to follow."[54]

Finding this middle way was not easy, and undoubtedly many men failed in the effort, as does Jacques Arnoux the china manufacturer in Flaubert's *L'Éducation sentimentale.* Flaubert was more cynical than Audiganne about the capacity of bourgeois manufacturers and consumers to appreciate art and taste. Explaining Arnoux's economic and social decline, the novelist describes the many disparate styles of china in the manufacturer's line of products. He implies that the reason Arnoux changes styles so frequently is that none of them sell well: "His intelligence was not high enough to attain to art, nor bourgeois enough to look merely to profit, so that, without satisfying anyone, he had ruined himself."[55]

---

53. Ibid.; Guillaumin, *Dictionnaire,* 379; Musée industriel, *Description complète de l'Exposition générale des produits de l'industrie française faite en 1834* (Paris: Société polytechnique et du recueil industriel, 1834) 3:40. Audiganne maintained that bronze making was a highly competitive industry, and that education in design was necessary if France was to continue to succeed in this area. Audiganne, *Industrie contemporaine,* 96.

54. Audiganne, *Industrie contemporaine,* 91.

55. Gustave Flaubert, *L'Education sentimentale* (1867), vol. 11 of *Les Oeuvres de Gustave Flaubert* (Lausanne: Editions Rencontre, 1965), 260.

The failure of the fictional Arnoux and the real success of Christofle and Barbedienne suggest that understanding consumer demand was critical for manufacturers in art industries. By themselves, new products and low prices were insufficient for attracting bourgeois customers in nineteenth-century France. Christofle, Barbedienne, and others also appealed to consumers' desire for the appearance of luxury and art in the home. This insight on the part of the manufacturers, as well as their capacity for innovation, merits consideration, especially in the context of scholarly reevaluations of industrialization. Electroplating and pantograph reproductions succeeded because they imitated the authentic article—solid gold or silver objects, or original statues—elegantly and accurately. No earlier methods of metal plating or reproduction were so effective. Because there was a growing bourgeoisie striving, on more or less limited incomes, to imitate the ruling elite, manufacturers in *orfèvrerie* and bronze making were able to expand their industries to include radically new methods of production and a new and larger market. Nonetheless, the products retained some characteristics of art; and this is why consumers bought them.[56]

The examples from the Crystal Palace exhibition discussed in this chapter indicate that French art manufacturers could indeed combine entrepreneurial abilities with high standards of quality and aesthetics to the benefit of French industry as a whole. But jury members and other men concerned about the future of French industry feared that after 1851 France might lose its market niche in art and luxury manufacturing. They prophesied that other countries would improve the design and quality of their products, and so diminish the foreign markets for French goods. Moreover, they anticipated that large-scale industries in France that were protected by high tariffs and prohibitions would continue to create difficulties for smaller enterprises, through more expensive raw materials and retaliatory tariffs on exports. The very success of men like the *orfèvres* and bronze makers at the Crystal Palace exhibition generated volumes of literature on how the government should encourage such manufacturers and so promote French industry in general. The suggestions of these policy makers and would-be policy makers are the subject of part 3.

56. Sylvie Forestier, "Art industriel et industrialisation de l'art: L'Exemple de la statuaire religieuse de Vendeuvre-sur-Barse," *Ethnologie française* n.s. 8, no. 2–3 (1978): 191–200.

Part Three

Taste in Politics
The Exhibition as a Watershed

Chapter Six

Art for Industry's Sake

Léon de Laborde's Plan for Transforming Taste

In 1850 the art critic Count Léon de Laborde (1807–69) despaired of French society, culture, and politics. With his aristocratic background and his record of distinguished service to the governments of the Restoration (1815–30) and the July Monarchy (1830–48), Laborde found himself at odds with the Second Republic (1848–52) that emerged from the 1848 revolution. He criticized the Republic as responsible for the physical destruction that accompanied the revolution, and he was pessimistic about the government's ability to maintain social order and peace. Even more frightening to Laborde were the nefarious implications of democratic republicanism for art. Pointing to the republics of Switzerland and the United States, he implied that France, too, under its current administration might experience "the annihilation of the arts by democratic jealousy, by a materialistic attitude, by mercantile preoccupations."[1] Laborde's beleaguered mentality was apparent in the introduction to his study of art under the kings of the French Renaissance; he recommended that lovers of art should withdraw from society and lead an almost underground existence, since art appreciation in such "disordered times" was at an all-time low.[2]

However, the Crystal Palace exhibition of 1851, and the establishment of the Second Empire a year later, transformed Laborde's self-pitying resignation to reforming zeal. Laborde had been a member of French industrial exhibition juries since 1839, and it was logical

---

1. Le Comte [Léon] de Laborde, *La Renaissance des arts à la cour de France: Études sur le seizième siècle* (1850; New York: Burt Franklin, 1965) 1:xxxii.
2. Ibid., xlvii.

that he be appointed head of the section on the fine arts applied to industry for the 1851 exhibition. In this capacity Laborde published in 1856 a thousand-page report, ostensibly analyzing fine arts at the Crystal Palace but in fact explaining how and why the arts had declined in France and what was necessary to reverse that trend. Curiously, this aristocrat steeped in the art of the European past found in the first international exhibition of industry the starting point and the justification for an ambitious package of cultural reforms that he proposed to the French government. This plan for massive state intervention in art reform was indeed better suited to the authoritarian, centralized, imperial nation-state of Louis-Napoleon Bonaparte than to the democratic, egalitarian, social republic of 1848.[3]

But why was Laborde so concerned about the decline of the arts in France in 1851? French manufacturers had won international acclaim for the artistic merit and good taste of their products at the Crystal Palace exhibition, and Laborde readily acknowledged France's artistic superiority over other industrializing countries. Yet he feared that this advantage was precarious in a modern, industrial age; and he was not alone in this regard. Reforming art instruction in professional schools and fine arts academies was an issue of serious debate and action in France under the July Monarchy.[4] In addition, German and English writers, along with the French, responded to the 1851 exhibition with several books and pamphlets on how industrializing countries should improve the artistic quality of their manufactured goods.[5] Among the distinguishing and significant features of Laborde's treatise, however, was his emphasis on consumer taste as the key to a nation's artistic success or inferiority.

Long before the theoretical critiques of consumerism of the late nineteenth and early twentieth century,[6] Laborde understood the

3. A provocative analysis of art under the Second Republic is T. J. Clark, *The Absolute Bourgeois: Artists and Politics in France, 1848–1851* (London: Thames and Hudson, 1973).

4. Albert Boime, "Entrepreneurial Patronage in Nineteenth-Century France," in *Enterprise and Entrepreneurs in Nineteenth- and Twentieth-Century France,* ed. Edward C. Carter III, Robert Forster, and Joseph N. Moody (Baltimore: Johns Hopkins University Press, 1976), 174–75.

5. Nikolaus Pevsner, *Academies of Art Past and Present* (Cambridge: Cambridge University Press, 1940), 248–56.

6. Rosalind H. Williams, *Dream Worlds: Mass Consumption in Late Nineteenth-Century France* (Berkeley: University of California Press, 1982). See also

fundamental importance of consumer demand in influencing the art and style of manufactured goods. His broad perspective on consumption, including an awareness of the political, social, and economic conditions of a given time and place, derived from an essentially conservative view of history combined with remarkable appreciation for certain nineteenth-century developments such as mechanized production and mass politics. Laborde associated France's artistic decline with the downfall of the monarchy, the democratization of consumption, and the end of craft guilds. But instead of looking backward to the restoration of the old order as a means of reviving French art, he called upon the modern nation state to use mass education, as well as other public institutions, to cultivate a nationwide aesthetic. Laborde was proposing more than simple government reform of art and professional education. In order to ensure that the French population bought only what he considered good art and aesthetically pleasing manufactured products, Laborde sought to dissolve any distinction between public art and private taste. In short, Laborde's plan entailed the use of state power and modern technology to create an aesthetic shared by consumers and producers alike.[7] How Laborde arrived at this position, with his views of art and history and the influence of the Crystal Palace exhibition, is the subject of this chapter.

AN ARISTOCRATIC PUBLIC SERVANT

Born in Paris during the reign of Napoleon I, Léon de Laborde was oriented early on to his father the count's interest in art and archaeology, and to his comparatively progressive social and political ideas.[8] The young Laborde studied at the University of Göttingen, an institution that was more innovative in its teaching of art than the

---

Thorstein Veblen, *The Theory of the Leisure Class* (New York: Macmillan Co., 1899) and Georg Simmel, *The Philosophy of Money,* trans. Tom Bottomore and David Frisby (London and Boston: Routledge and Kegan Paul, 1978).

7. Walter Benjamin, "The Work of Art in the Age of Mechanical Reproduction," in *Illuminations,* trans. Harry Zohn, ed. Hannah Arendt (New York: Harcourt, Brace and World, 1968), 219–53.

8. Dr. Hoefer, *Nouvelle Biographie générale* (Paris: Firmin Didot frères, 1858) 27:385–88. In the exhibition report, Laborde refers proudly but not uncritically to his father's proposal of 1815 for extending public education to all social classes. Whereas Laborde *père* advocated the teaching of reading and writing to everyone, reserving art education for only the wealthy elites, his son proposed both literacy and art instruction for all (see the discussion later in this chapter).

French art academies. He then went with his father to the Middle East to study and draw the remains of ancient civilizations. From 1828 to 1836 the young man held several diplomatic posts, until he decided to devote himself more fully to literature and art. Laborde published several works on a variety of topics, including travel in the Middle East, the history of printing, urban planning in Paris, art and court life in medieval Burgundy, and art and court life in Renaissance France. During the July Monarchy he also began serving regularly on the juries of national industrial exhibitions.

Around the time of Laborde *père*'s death in 1842, the son assumed further responsibilities and posts. He was elected deputy to the National Assembly in 1840 and in 1846, replacing his father as representative from Seine-et-Oise. In 1842 he succeeded the older Laborde as a member of the Institut, in the Académie des Inscriptions et de la Littérature. Laborde became curator of ancient art at the Louvre in 1845; he lost the position in the revolutionary year of 1848, but two years later was reinstated at the Louvre, this time as curator of the medieval and Renaissance collections.[9] That same year the minister of agriculture and commerce, Louis-Joseph Buffet (1818–98), appointed Laborde to the jury for the upcoming international exhibition of industry in London, and Laborde immediately found himself embroiled in a controversy over the British committee's decision regarding fine arts at the exhibition.

The men planning the exhibition decided to admit sculpture and architecture, but they excluded painting on the grounds that the latter, "being but little affected by material conditions, . . . seemed to rank as an independent art."[10] Protesting the exclusion of the one area where French exhibitors were certain to succeed, the Académie des Beaux-Arts called upon its members and all artists to boycott the exhibition. Though the British committee offered to construct a separate building solely for the exhibition of painting alongside the Crystal Palace, the French government refused to pay the transportation costs of shipping paintings to London, along with the exhibits

9. Ibid., 388–90.
10. *Reports by the Juries on the Subjects in the Thirty Classes into which the Exhibition was Divided* (1852), 1547, quoted in Patricia Mainardi, *Art and Politics of the Second Empire: The Universal Expositions of 1855 and 1867* (New Haven: Yale University Press, 1987), 25.

of industry for which money had already been allotted. Without French exhibitors, the plan for a separate painting display collapsed.[11]

This tussle over the distinction between art and industry was anathema to Laborde. Castigating the academy's position, Laborde urged French sculptors to display their works in the Crystal Palace as part of the fine arts section. Though he succeeded in persuading some of these artists to participate, Laborde was sorely disappointed that French sculpture was not represented in the numbers and quality that it could have been.[12] The preexhibition controversy only confirmed Laborde in his belief that separating art and industry was harmful to France's stature as an industrial power. This belief underlay Laborde's entire analysis of the exhibition, of the decline of French art, and of the means of artistic renaissance.

LESSONS FROM THE EXHIBITION

The overarching theme of Laborde's lengthy report on fine arts at the exhibition was the unity of art and industry.[13] He explained how the separation of art and industry had occurred in French history, why this development was so terrible in view of the tasteless and ugly products manufactured in his own day, and finally in what manner the state should reunify art and industry in the future. The Crystal Palace exhibition, allowing for the comparison of manufactured goods from all over the world, provided Laborde with much of the evidence he needed to make his case for artistic and industrial unification. Particularly instructive for Laborde, as for most French observers of the exhibition, was the comparison of products from Britain and France.

In Britain more than in France, Laborde contended, manufacturers exploited scientific discoveries and technological innovations to bring down the costs of production and to lower the price of manufactured

11. Mainardi, *Art and Politics,* 25–30.

12. Commission française sur l'Industrie des Nations, *Exposition universelle de 1851: Travaux de la Commission française sur l'Industrie des Nations* (Paris: Imprimerie impériale, 1856) 8:234–38.

13. Indeed, in 1856, the year that the imperial press published Laborde's treatise as the eighth and final volume of the official jury report on the exhibition, it also published the identical work as a two-volume book under the title *De l'union des arts et de l'industrie.*

goods. He praised highly these benefits of scientific and technological development, but the performance at the exhibition indicated that the British had neglected the art and quality of products in their pursuit of lower costs. French superiority in matters of art and taste, Laborde continued, was due to their giving priority to artistic merit rather than low prices in manufactured goods. However, both characteristics were essential for economic success; Laborde proclaimed that, starting in 1851, "it was understood that the sciences and the arts were the two inexhaustible breasts from which industry takes its nourishment and renews its energy."[14] Earlier in his report, Laborde had suggested that art and industry since the French Revolution comprised two (artificially) separate but equal aspects of human civilization; but here he claimed that industry was dependent on both scientific (including technological) and artistic developments. The metaphor worked well enough for Laborde's diagnosis of and prognosis for British industry, based on the exhibition and its immediate aftermath.

It was obvious to the British, as to everyone else, that they had sacrificed art for low price in manufactured goods. According to Laborde, the exhibition had revealed to all that art was the secret of French industrial success. He added that in response to this discovery, industrial countries were reforming art education at home to compete more effectively against their French rivals in the future. Laborde was most concerned that the British would actually succeed in improving the artistic quality of their manufactured goods, due to the establishment of a new school and museum of applied art, the government's purchase of many new models for art instruction (including several exhibits from the Crystal Palace), and Britain's already highly advanced technology.[15]

If Britain responded to the exhibition with art reform in order to balance the influence of technology and art on British industry, then logically France should have promoted technological innovations to make French products cheaper as well as artistic. From Laborde's perspective, however, developing French technology was not the only nor the best means for France to compete internationally, though he consistently supported it as intrinsically worthwhile. He wanted France to maintain artistic superiority, not just to become the

14. Commission française, *Exposition* 8:385.
15. Ibid., 382–91.

industrial equal of England. Thus Laborde's conclusion from the exhibition was that France, too, must improve artistic quality in industrial products if it was not to lose its advantage to Britain. Building art schools and museums was a good way to start, and the French could certainly learn from the British efforts at art reform. But Laborde had a much more ambitious project in mind to uphold France's industrial might through its artistic superiority—a project that required the financial and ideological backing of the French government.

Laborde's conviction that only the government could effect the kind of art reform he deemed necessary derived from the huge scope of his reform plan and from his view of art history in France. He believed that the ultimate target of successful art reform was the consumer; and since all the inhabitants of France were consumers, the reforms had to reach every man, woman, and child. "Industry depends on the public," Laborde wrote, so that the quality of manufactured products was only as good as the public demanded.[16] To cultivate good taste and art appreciation among the masses, Laborde proposed unprecedented state intervention in education, manufacturing, and public works. But how to induce the Second Empire to foot the bill for these huge expenditures?

Laborde was not above a little flattery directed toward the emperor, and a judicious invocation of history. Citing Cosimo de' Medici, Pope Leo X, Francis I, and Louis XIV, Laborde suggested that Louis-Napoleon Bonaparte, too, could achieve immortality by initiating an artistic renaissance in nineteenth-century France. "It belongs to memorable eras, to able sovereigns, to great nations, to be distinguished by works of art," Laborde asserted.[17] He went on to intimate, in yet another bodily metaphor, that through the state's promotion of the arts the emperor could make France the lungs, and Paris the heart, of the chest of Europe.[18]

Laborde believed that past French kings and aristocrats were responsible for his country's international artistic leadership, and his goal was to mobilize the modern nation-state of the Second Empire to fulfill that same function in the contemporary context. Favorable

---

16. Ibid., 431. Cf. Williams's analysis of Camille Mauclair's decorative arts reform movement of the 1890s in Williams, *Dream Worlds,* 162–67, 180–85.

17. Commission française, *Exposition* 8:444.

18. Ibid.

to the absolutism in French history, Laborde was untroubled by authoritarianism in his own time; in this regard, among others, he differed significantly from the leaders of art reform under the Third Republic (1870–1940).[19] Whereas his successors sought symbols and styles to represent and reinforce the principle of individual liberty in bourgeois democracy, Laborde wished to see public institutions impose an elitist aesthetic on the masses. An explanation for this apparent paradox must begin with Laborde's interpretation of art and history.

LABORDE'S HISTORY OF ART AND INDUSTRY

The touchstone of Laborde's history was ancient Greece. According to him, the classical style was and remained the pinnacle in the history of art because of its purity and harmony. Underlying this aesthetic triumph, in Laborde's view, was the fact that all Greek citizens shared the same standards of beauty and taste. Through art education, urban architecture, and public ceremonies like the Olympic games, the leaders of ancient Greece ensured universal appreciation and perpetuation of the same style in all objects, from household utensils to religious temples. Laborde asserted that a wonderful consequence of the uniformity of style was the lack of distinction between artists and manufacturers. "Among the Greeks there were neither artists nor industrialists, but there were artists of differing degrees of talent, whose productions were distinguished especially by their difference of purpose [*destination*]."[20]

While Laborde's interpretation of social and cultural conditions in ancient Greece is open to question, it was important for his understanding of the subsequent history of art in Europe, and for his proposals to rejuvenate art in his own time.[21] In essence, the remain-

19. Miriam R. Levin, *Republican Art and Ideology in Late Nineteenth-Century France* (Ann Arbor: UMI Research Press, 1986); Debora L. Silverman, *Art Nouveau in Fin-de-Siècle France: Politics, Psychology, and Style* (Berkeley: University of California Press, 1989).

20. Commission française, *Exposition* 8:15.

21. In his praise for the uniformity of public taste among the Greeks, Laborde ignored the division between citizen and slave in ancient society. In addition, Laborde's portrayal of women in ancient Greece resembles nothing so much as the nineteenth-century domestic ideal and separation of spheres. Laborde was oblivious to class and gender differences that might have disturbed his view of perfect social and cultural harmony. Ibid., 7–23.

der of Laborde's historical account addressed the extent to which later societies and nations were able to approximate the aesthetic stability and the union of art and industry that characterized his version of the ancient Greeks. Laborde concentrated on the history of France, and he praised several French kings for their patronage of the arts and for their encouragement of distinctive styles, notably Francis I for the development of a French Renaissance style inspired by the Italian.[22] However, Laborde also chastised these royal connoisseurs for contributing to the separation of art and industry. This process was linked with the establishment of guilds, especially in the thirteenth century, and the exemption of particularly gifted artisans from guild regulations so that they could produce innovative works of art for princes both secular and ecclesiastical. For Laborde, this marked the beginning of a hierarchy whereby court artisans considered themselves artists because they created for the king, and guild artisans were merely craftsmen who produced for lesser folk.[23]

In Laborde's history a laudable effort to encourage and regulate the arts was the organization of state manufactures by Louis XIV's minister of finance Jean-Baptiste Colbert (1619–83) in the seventeenth century. Though art under Louis XIV lacked originality because of the excesses of centralized authority in the state manufactures, Laborde welcomed the monarch's creation of aesthetic "stability" out of chaos. It was Louis XIV's absolute power that allowed for these artistic reforms, which in turn enhanced the king's influence domestically and abroad. In an absolutist society, public opinion hardly mattered, and Laborde indicated that the masses never saw any of the fine art and furnishing that Louis XIV collected and commissioned. As in his treatment of ancient Greece, however, Laborde presumed to know the popular mentality, and he asserted that the common people of the seventeenth century, in their ignorance of bad art and taste, appreciated the public manifestations of Louis XIV style. "The people of that time loved the arts and felt them

22. Ibid., 58–97.
23. Ibid., 49. One interpretation of this phenomenon, based on the study of seventeenth- and eighteenth-century court artisans, argues that their exemption from the guilds forced them to operate in essentially a laissez-faire market, with all its freedoms and insecurities. Laborde, as we shall see, considered the free market harmful to art unless the state controlled consumer tastes. Michael Sturmer, "An Economy of Delight: Court Artisans of the Eighteenth Century," *Business History Review* 53 (Winter 1979): 496–528.

better than people do today, because their taste had not been led astray by the bad pastiches presented in the last fifty years. . . . Art, for them, consisted of the monumental application of its beauties and splendors."[24]

The arts in French history enjoyed both brilliant epochs and periods of dismal obscurity, but the high points only existed because of royal leadership and state intervention, according to Laborde: "In all eras, from Charlemagne to Louis-Philippe, our kings were responsible for fruitful protection [of], and almost always enlightened influence [upon, the arts]."[25] An ignorant public, such as that of the seventeenth century, was simply incapable of generating artistic revivals or creating a distinctive style. For Laborde, only powerful rulers could adequately support the arts, establish new styles, and impose aesthetic values upon the masses. He acknowledged that monarchs could create styles of dubious quality, such as rococo under Louis XV; but he always favored any style, or what he called aesthetic stability, over no style at all ("chaos"). It comes as no surprise, then, to learn that in Laborde's interpretation the French Revolution of 1789 was disastrous for art and inaugurated a sixty-year decline that he saw reflected even in the finest specimens of French products at the Crystal Palace exhibition.

Laborde enumerated several consequences of the French Revolution that threatened to undermine a national artistic ascendancy established over several centuries of elite patronage. For one thing, by effectively eliminating the aristocracy the French Revolution destroyed the only social group with the wealth, and especially the cultivation, to patronize the arts successfully. Another problem Laborde cited was the end of guilds, the only popular institution that had retained any semblance of the unification of art and industry through the transmission of skills and some appreciation for art. For Laborde, however, the outcome of the French Revolution that was most detrimental to the arts and to aesthetic stability was the free market economy that made mass consumption and especially bourgeois patronage the arbiters of taste in France.[26]

Ever since 1789–99, according to Laborde, artists and manufacturers had to produce what the public wanted, not what was aesthet-

24. Commission française, *Exposition* 8:112.
25. Laborde, *Renaissance,* i.
26. Commission française, *Exposition* 8:197–200.

ically meritorious. Laborde excoriated popular taste for adopting the English value of comfort, and for abandoning classical style in favor of excessive romanticism. He claimed that wealthy bourgeois, who replaced the aristocracy as purchasers of art, were interested in art only as a form of speculation, and so made mere merchandise out of what Laborde considered to be the philosophical and moral manifestation of civilization. No distinctive style had emerged in France for the entire first half of the nineteenth century; Laborde blamed this failure on the ignorance and capriciousness of the consuming public and on the inability of political leaders to appoint "men of taste" to head artistic and industrial institutions.[27] Though France won plaudits at the Crystal Palace exhibition for the artistic quality and good taste of its manufactured goods compared to those of other industrializing countries, Laborde believed that this experience only aggravated the problem of French people's exaggerated confidence in their artistic superiority. For Laborde the Crystal Palace exhibition represented a turning point in the tumultuous history of art in France. It revealed to the entire world a profound aesthetic crisis, one that, according to Laborde's interpretation of history, only strong state leadership could rectify.

As the centralized state had established French artistic ascendancy in the past, so Laborde turned to it again for leadership in the nineteenth century. But Laborde was enough of a realist to understand that merely returning to royal absolutism or aristocratic domination would not serve his purposes. The modern French state of the 1850s, based nominally, if not in practice, upon popular support, could not ignore public opinion in either political or cultural matters. Thus Laborde called upon the government to use the financial and administrative power at its disposal to indoctrinate the public with uniform, and what Laborde viewed as universal, aesthetic principles. Education was the primary means Laborde advocated for reviving art and unifying art and industry in France. He also proposed the establishment of a huge new state manufacture of consumer goods, as well as other public institutions of art and industry. Laborde was convinced that modern means could successfully restore traditional values; in this case, mass education would disseminate an elite aesthetic.

27. Ibid., 203–11.

DRAWING FOR EVERYONE

Laborde had particular ideas about art instruction for workers and artists, but his educational reforms all rested upon the teaching of drawing as part of the curriculum for all pupils throughout their education.[28] He proposed that drawing should be taught simultaneously with writing; he considered drawing to be a skill, like writing (and horseback riding, ice skating, and swimming), that may diminish for lack of use, but that an individual never totally forgot when he or she had learned the ability in infancy. Laborde therefore suggested that from the very earliest years of life, all children should learn drawing from their mothers or in day-care centers, which he asserted should be more numerous and better equipped. Citing the educational reformer Friedreich Froebel, Laborde advocated visual stimuli and artistic toys as essential to the successful teaching of drawing to children, and he counseled that the walls of kindergartens should be decorated with pictures of the Egyptian pyramids, the Greek Parthenon, and the Roman Colosseum. Indeed all schools, and implicitly all private homes, should ideally contain examples of the finest art in the history of the West, and especially the art of the ancient Greeks, to contribute to the cultivation of the arts and industry in France.[29]

According to Laborde, children two to six years old should be taught drawing as a game rather than as a discipline. From the ages of six through ten children should practice drawing as a manual skill; from ten to sixteen they should be encouraged to draw with artistic feeling and should learn the principles of proportion. Just as the process of learning to draw must vary with children's ages, so, too, should drawing exercises be presented to children on an ascending scale of complexity and difficulty. The first phase, according to Laborde, was to require that children draw directly from models; the state must assume responsibility for engaging the best contemporary artists to create models of the human form and copies of great sculptures for art instruction. Such a steady demand for models, Laborde anticipated, would stimulate manufacturers to compete with

28. Third Republic plans for art reform also included drawing as part of the curriculum to educate both producers and consumers. Levin, *Republican Art*, 79–87.

29. Commission française, *Exposition* 8:512–13, 569, 571, 593–95. Earlier in his essay Laborde deplored the dullness of the prestigious Lycée Louis-le-Grand, where he sent his own son to school. There were no pictures on the wall there.

one another in developing the most accurate methods of reproducing models at the lowest cost. The second phase in drawing instruction was to draw from memory; finally, the most difficult task would be tackled, original composition.[30]

And who was to teach drawing to the boys and girls of France, once they left home for either public or private schools? Laborde, of course, had the answer: a uniformly trained, well-paid corps of drawing masters. They would all use the same instruction manual and the same methods of teaching. "There is only one art, only one method for teaching it, only one ideal of beauty to understand," Laborde declared.[31] The most important attributes of a good drawing master for children, from Laborde's perspective, were authority and power. Secondarily, masters should also understand the individual child's character, and so more successfully coax their young pupils to achievement in drawing. Laborde was well aware that his plan required incentives to attract competent persons to the career of drawing master; for this purpose he proposed large salaries, teacher and student competitions, and regular exhibitions of teachers' drawings with prizes awarded to those that showed talent.[32]

WOMEN AND ART

It is clear from his emphasis on infant education that Laborde credited women with a significant role in the elevation of the arts in France. As the first instructors of young children, and the decorators of the home, mothers were essential to the success of Laborde's plan. A woman, properly educated, was for Laborde the potential provider of "a school in the family," and he was certain that once the public schools included drawing instruction in the curriculum for both boys and girls, parents would see the advantage of beginning art education in the home to help their children achieve success at school.[33] But women were to perform another, even more important function in Laborde's modern Renaissance; he believed that with instruction in art women could contribute to French industrial strength and to social order.

30. Ibid., 572, 618–23.
31. Ibid., 579.
32. Ibid., 582–92.
33. Ibid., 523.

Laborde, like many of his contemporaries, believed that women more than men were especially inclined toward "minor" artistic abilities.[34] "[In the past, woman] possessed to a higher degree than [man] the intelligence that swiftly assesses the most seductive aspect of things, developed by the coquetry of her dress, the skillful hands that execute delicately, and a highly developed faculty to understand and reproduce all that has charm and grace."[35] Laborde argued that in his own day, women had lost this particular advantage over men because they, much more than men, were deprived of education in art. He invoked the dismal situation of women of the working class, as well as the need for French industrial supremacy, in support of his proposal to educate women in art. Referring to massive unemployment among women due to the use of machine technology in the textile industries, Laborde suggested that the solution to this problem was to endow women with artistic knowledge and skills. Occupations like porcelain painting, for instance, that required aesthetic understanding as well as manual dexterity, would never be replaced by crude machines. Moreover, he deemed artistic occupations particularly suitable to women because they could work at home. "They paint, sculpt, engrave, make lithographs, design embroidery and fans, and produce a thousand delicate art objects called *articles de Paris* without leaving the mother's protective roof, without losing sight of the baby's cradle, without placing a foot in these places of corruption called factories, veritable prisons of communal labor."[36] Laborde thus viewed artistic occupations for women as a means of reinforcing the feminine ideal while simultaneously promoting industrial development.[37]

For economic, moral, and social reasons, then, educating females in art was just as important as compulsory art instruction for males.

34. See chapter 2.

35. Commission française, *Exposition* 8:520.

36. Ibid., 523.

37. Several male politicians and theoreticians advocated home work for women as more moral than working outside the home. For example, Charles Dupin, *Le Petit Producteur français*, vol. 6, *L'Ouvrière française* (Paris: Bachelier, 1828); Jules Simon, *L'Ouvrière* (Paris: Hachette, 1861). See also Joan W. Scott, "Statistical Representations of Work: The Politics of the Chamber of Commerce's *Statistique de l'Industrie à Paris, 1847–48*," in *Work in France: Representations, Meaning, Organization, and Practice*, ed. Steven Laurence Kaplan and Cynthia J. Koepp (Ithaca, N.Y.: Cornell University Press, 1986), 335–63.

To be sure, Laborde admitted some differences in the art education for boys and girls, since he was certain that females in nineteenth-century France would find Greek statues of male nudes offensive. While Laborde personally thought that such sculptures were beautiful and not at all immoral, he bowed to contemporary mores and suggested substitutes for the Farnese Hercules and the Belvedere Apollo as models for female art pupils to study and draw.[38] Despite his proposal for more or less equal education in the arts for males and females, Laborde by no means advocated a change in gender relations in French society. Though education and training would, in principle, grant women greater job security and presumably better jobs with higher pay, Laborde obviously hoped that the "new" female occupations would not interfere with household and child-care responsibilities. Laborde ignored the possibility of female artist workers competing with men, probably because he expected that women's paid work would be subordinated to the home. Moreover, he believed that the advancement of civilization entailed increased demand for beautiful, artistic, handmade or hand-decorated goods—just the kind of products that women trained in art would make. In an ideal society (and Laborde definitely thought he was contributing to the attainment of one) there would be plenty of jobs for all workers, female and male alike. But workers, like women, required instruction in drawing and design to participate in this brave new world.

ART FOR WORKERS

Anticipating the criticism that education in art would disrupt the social hierarchy, would instill in the working class discontent with their lot and a wish for higher status, Laborde insisted that his reforms would actually render workers more satisfied with their jobs. He assured his readers that workers' appreciation for the aesthetic quality of a finished product would overcome their alienation due to division of labor. Even the lowliest worker, performing the most basic of preparatory tasks in the production process, would be aware of his contribution to the final outcome. Moreover, the increasing use of steam power and mechanized production would relieve workers of the most strenuous and mind-numbing occupations, freeing them for

38. Commission française, *Exposition* 8:524.

more satisfying jobs that called upon their brains instead of their brawn.[39]

Additionally Laborde argued that art education for workers was necessary to enhance France's status as an industrial nation. According to Laborde, the artistic shortcomings of French manufactured goods at the Crystal Palace exhibition were mainly due to the inability of French workers to design furniture and furnishings appropriate to their use. Laborde credited French workers with great manual skill, but with no understanding of art. Art education, he contended, would help workers appreciate that a product's design, raw materials, and function all had to be accounted for to create a work of art. In the absence of the guilds, destroyed by the French Revolution, workers needed some other means of learning their trades, and especially of understanding the fundamental unity of form and function.[40] Drawing schools, strategically located in the capital, were part of Laborde's proposed alternative to craft guilds.[41]

On the basis of his calculations of the number of workers residing in the different arrondissements of Paris, Laborde contended that the state must erect four drawing schools to serve a student body of 20,000 to 25,000 workers. He estimated class sizes of 200 pupils, and he was confident that a competent instructor, assisted by monitors, could successfully teach drawing to such a large number. While the main purpose of these free, public schools was to graduate artist workers, they also were expected to cultivate diligence and regular work habits among pupils. "[Apprentices and workers] will be admitted at no charge, and medallions of some value will be distributed among pupils on Sunday and Monday evenings. This latter measure is intended to make pupils prefer the class to the cabaret."[42]

Laborde hoped that these drawing schools would contribute to both social and aesthetic unity. Though concerned about the moral weakness of workers, he believed that art education would improve relations between manufacturers and their employees, by fostering near equality. This equality would rest on mutual respect for each

39. Ibid., 495–96, 506, 651.

40. Ibid., 636–43.

41. Laborde explained his focus on Paris by stating that it had always been the center of art and industry in France, and that the benefits of the schools in the capital would soon spread to the provinces. Ibid., 651–55.

42. Ibid., 662. Cf. the radical reform of art education for workers described in Boime, "Entrepreneurial Patronage," 174–75.

other's artistic knowledge and skills—with Laborde's proposed reforms, everyone would share a solid background in drawing and
aesthetics. Workers would be elevated to artists through education,
and employers would relate to them as equals because "they speak the
same language."[43] In a society where everyone was an artist, class
conflict would have no place.[44]

Similarly, Laborde's professional schools were supposed to help
eliminate the distinction between "applied arts" and "fine arts." When
all French workers learned drawing and studied ancient Greek models, the aesthetic that was manifested in painting, sculpture, and
architecture would also inform furniture, fabric, pottery, jewelry,
and so forth. But even Laborde could not escape from linguistic and
conceptual distinctions between workers and artists, decorative arts
and fine arts. He admitted that workers had to learn specific rules
about the application of art to ornamentation that differed from the
rules learned by future painters, sculptors, and architects. "The artist
worker . . . knows that nature does not provide ornamentation, that
she presents to man only motifs in inexhaustible variety, and that
man must modify the model according to circumstances."[45] Workers, then, had to learn that design or decoration must be appropriate
to the raw material and especially to the function of the product.
These constraints did not apply to the students who would be artists,
as opposed to artist workers. Indeed, Laborde's notion of education
in the fine arts differed radically from his proposals for the masses and
for workers, confirming the aesthetic and social distinctions that he
was trying to eliminate.

EDUCATING ARTISTIC ELITES

For talented young people who wished to become painters, sculptors,
and architects, Laborde proposed a third and final program of art
education that differed pedagogically from art instruction for children

43. Commission française, *Exposition* 8:649–50.

44. Laborde never indicated exactly where entrepreneurs would receive their
art eduction beyond the elementary level; the implication is that they would
come from the ranks of the "apprentices and workers" in the professional
schools.

45. Commission française, *Exposition* 8:670–71. On the distinction between
fine art and decorative art, see Charles Laboulaye, *Essai sur l'art industriel* (Paris:
Bureau du dictionnaire des arts et manufactures, 1856), 6.

and workers. Whereas Laborde valued obedience and diligence among art pupils in elementary and professional schools, he extolled rebelliousness, independence, even laziness on the part of fine-arts students. One of the few positive results of the French Revolution for Laborde was, surprisingly, the subversive attitude of youth: "The revolution of '89, in overthrowing the hierarchy of the previous society, produced a new youth, which rebels against all submission, and wants to hear about neither the slow pace of achieving its rights nor progress in the enjoyment of independence; it requires every-thing, and immediately."[46] Among art students this spirit manifested itself in arrogant disregard of professors, and a preference for work-ing without the masters' guidance. While Laborde was critical of this practice, he discerned in art students' rejection of authority the po-tential for artistic creativity. He called upon master artists not to punish these students, but to cultivate their originality. In order to become professors, experienced artists had to win students' attention and willingness to learn through their own passion for and commit-ment to art. According to Laborde, the task of the fine-arts professor was to encourage students to discover on their own, and only inter-vene to prevent students from imitating too assiduously other artists and styles.[47]

Liberal as this position was, particularly in the context of the inflexible and highly imitative method of teaching fine arts in La-borde's time, there were limits as to what Laborde considered "orig-inality." Portraying ordinary scenes and animals was not art and never could be, as far as Laborde was concerned. He abhorred the principle of art for art's sake, interpreting it as a justification for aesthetic anarchy: "Not long ago it was possible to assert seriously that a cabbage, perfectly painted, must be considered the equal of Raphael's *Transfiguration*. That was the triumph of the bourgeoisie, of the banalities of the day, of the reality of life."[48] For Laborde only historical and mythical subjects could truly be art. Studying nature was an important part of an artist's education, but not an end in itself.

46. Commission française, *Exposition* 8:678.
47. Ibid., 679–83.
48. Ibid., 737. Laborde was clearly not a friend of the realist movement of the mid-nineteenth century. Linda Nochlin, *Realism* (New York and Baltimore: Penguin Books, 1971).

Largely on the basis of Laborde's proposed reforms of fine-arts education, art historians consider him a liberal and progressive thinker. Rejecting the stifling copying and excessive attention to detail that prevailed in the Academy of Fine Arts in the middle of the nineteenth century, Laborde emphasized originality, the large effect, and even the potential strength of an art student's weakness.[49] But to characterize Laborde as a liberal reformer is to underestimate the full scope and implications of his project. Regarding education alone, and ignoring for the moment his schemes to homogenize consumer tastes, Laborde's plans included decidedly illiberal elements. For one thing, his aesthetic was narrow and rigid; ancient Greek art was its basis, and only certain subjects, containing the human form in historical or mythical contexts, could ever truly be works of art. To be sure, Laborde was exceptionally broad-minded in appreciating the overlapping of different artistic media. And in proposing the very same art education for all French children—male and female, rich and poor—Laborde's reforms were indeed democratic, or at least egalitarian. But the uniformity of drawing instruction—every master teaching the same material and in the same manner—is reminiscent of Napoleon's highly centralized and controlled higher-education program. Where Laborde encouraged the artistic elite in France to be bold and creative, conformity and submission to authority were the watchwords for the workers. Aesthetic and cultural unity were Laborde's goals, and his historical models for such unity were hierarchical, elitist, and absolutist societies. Small wonder, then, that Laborde's proposed educational reforms represented more the expansion of state power than the broadening of popular democracy.

Mass education, however, was only a part of Laborde's plan to effect an artistic renaissance in France and to unify art and industry. The other endeavor he proposed to the French government was what

---

49. Albert Boime, "The Teaching Reforms of 1863 and the Origins of Modernism in France," *The Art Quarterly*, n.s., 1 (Autumn 1977): 1–39; Albert Boime, "The Teaching of Fine Arts and the Avant-Garde in France during the Second Half of the Nineteenth Century," *Arts Magazine* 60 (December 1985): 46–57; Pevsner, *Academies*, 250. Many of Laborde's principles and suggestions were incorporated into an imperial decree of 1863 that drastically reformed fine-arts education amidst widespread controversy. Boime, "Teaching Reforms."

he called "the maintenance of public taste," involving nothing less than the complete absorption of individual taste to a public aesthetic.

TRANSFORMING CONSUMER TASTES

While Laborde grudgingly admitted that the French Revolution engendered two useful consequences—rebellious and creative youth, and free and equal education—he continued to bemoan the democratization of consumption that it also entailed. Everyone in Laborde's time could buy clothes, household furnishings, and even art in some form of reproduction. And in a free market economy (on a national, if not international, scale), all manner of consumer goods were available for private purchase. This limitless scope for individual choice was abhorrent to Laborde; for him, it was the cause of ugly manufactured products and of general artistic anarchy. According to Laborde, ignorant and increasingly numerous consumers tyrannized over manufacturers and artists, forcing them to produce works in the latest fashion at the expense of artistic quality: "The manufacturer is an echo who responds to the good or bad taste of consumers."[50] Although technological and scientific developments in industry contributed to the growing array of eclectic and tasteless manufactured products, Laborde identified the problem as primarily one of consumer demand rather than product supply. "As for this liberty, which is abandonment, which is laissez-faire, I know of no more odious tyranny, for it places the artist entirely at the mercy of the public."[51]

For Laborde the "tyranny of the masses" in modern French society was far more detrimental to art than elite tyranny in ancient Greece or absolutist France. His underlying assumption was that elites in the past consumed for the public good, whereas contemporary consumers acted merely out of personal greed or whim. In Laborde's interpretation of history eras of great artistic achievement occurred when rulers successfully imposed their tastes upon an entire population. He adopted a patronizing attitude toward the masses who lived during such times, asserting that their very naïveté and childish curiosity made them receptive to the art manifested in the cathedrals and monuments that prelates and kings erected. "The people are fundamentally artists due to their naïveté, their facile credulity, their quick

50. Commission française, *Exposition* 8:432.
51. Ibid., 446.

enthusiasm."[52] For Laborde, the artistic sensibility of the people was corrupted in the aftermath of the French Revolution, when supposedly cultivated aristocrats no longer exercised the same influence over public taste. Consumption, in expanding to all social classes, became private, individual, and varied. Whereas in the past the state dominated art and public taste through its exclusive monopoly over wealth, in the nineteenth century too many people had access to money and to a huge array of goods for any single aesthetic or style to prevail. Rather than celebrating this new avenue for freedom and self-expression, Laborde sought to control it. He wished to exploit the extensive powers of the modern state to curb innovative consumption and impose a uniform aesthetic from above.

According to Laborde an important responsibility of the state was to support the arts and artists through plenty of commissions for public works and through financial aid to artists. In addition, the state should beneficently influence public taste through the sponsoring of libraries, athletic events, and parks.[53] Such public institutions, along with monuments and other works of art commissioned by the state and put on display, would confront people daily and in all ordinary activities with a common aesthetic. Significantly, Laborde himself proudly asserted that through these means, the distinction between private inclination and public art would disappear: "The entire nation, seeing the arts thus descend to the streets and mingle in public life, will understand that private life must in turn saturate itself with their gentle influence."[54] Although Laborde never referred to any particular style or symbolism that would represent the French state, he intended that state-sanctioned art should also become popular art through imposition and lack of alternative.

Finally, Laborde devoted hundreds of pages of his treatise to the topic of how the state should direct industry in France in order to uphold his aesthetic and to ensure French economic success in world markets. Laborde's primary means of accomplishing this objective was through the creation of a huge state manufacture, far more extensive than the existing enterprises of Sèvres, Gobelins, and Beauvais. Laborde considered these existing manufactures inadequate to the task of improving public taste and providing models for private

52. Ibid., 480.
53. Ibid., 804–5.
54. Ibid., 806.

industry. Only a huge state enterprise producing a complete range of consumer goods could revitalize French industry and lead it in a new and successful direction. Private industries, Laborde believed, had no incentive to change; they were locked into dependency on consumer tastes, and individually they lacked the capital for major reform. Thus, Laborde concluded, state intervention was absolutely necessary. His plans for this massive venture were detailed and extensive. Laborde had even chosen the site for this state manufacture in Paris — the former Île Louvier, comprising 30,000 square meters of mostly uninhabited land and including the resources necessary for manufacturing, such as water supply, good communication, and access to transport. In addition, Laborde enumerated the characteristics and qualifications of the ideal director of this public works — a true Renaissance man with proven abilities in drawing, sculpture, and architecture; a man of innate and impeccable taste who had traveled widely and who would liberally encourage the development of the arts.[55] Not coincidentally, Laborde himself met all of these requirements for administering such a vast, multifaceted enterprise.

The state manufacture that Laborde proposed would produce almost all durable consumer goods — carpets, tapestries, furniture, china, glass, crystal, jewelry, bronzes, gold and silver tableware, books, carriages, *articles de Paris,* and more. Rather than competing with private industries, the state manufacture would serve as a model to them, setting styles, inventing new technologies, training workers, and displaying products in public buildings for consumers' delectation and instruction. The government should furnish all public buildings completely with products from the state manufacture, and it should establish its own museum for a more permanent display of these goods. Laborde intended that the state manufacture would constitute a modern substitute for the old craft guilds, instructing apprentices and establishing a hierarchy of masters and workers. While this institution would preserve traditional techniques of manufacturing, it would also encourage invention; and Laborde envisioned for the manufacture's physical plant fully electric lighting, push-button windows, and swinging glass doors that opened and closed automatically. Mechanical methods of reproduction would no longer inundate the market with cheap, tasteless goods once the state

55. Ibid., 806–18.

manufacture was established. Instead, consumers would learn the value of well-made, artistic products, and demand only the best.[56] "Tastes in home decorations will become more discriminating, cheapness itself will seem to be a costly expense. . . . People will agree to pay more in order to possess something original, custom-made according to [the consumer's] tastes and appropriate for the space, height, and lighting of a particular apartment."[57]

Whereas most other commentators viewed the French showing at the Crystal Palace exhibition as a confirmation of the good taste and artistic sensibility of French consumers, Laborde treated it as the beginning of a state-sponsored process to achieve that end. Laborde's standards of what constituted art and what was aesthetically meritorious were obviously very high. He could not be satisfied, as others were, with French art that was merely superior to that of other countries. Moreover, by emphasizing the deficiencies in French manufactured goods Laborde lent urgency to his case for reforming art education and public taste in France as a means toward industrial competitiveness. The emphasis on improving French industry and competitiveness through such thorough state intervention in the arts was unique among French art reformers responding to the Crystal Palace exhibition. Although others shared Laborde's concern for artistic quality in manufactured goods, they lacked his broad social and political vision regarding consumption and production. Acting upon narrower professional interests, artists and entrepreneurs in France also appealed to the government to reform art and elevate the status of those engaged in the manufacture of decorative arts.

AN ALTERNATIVE VIEW ON ART AND INDUSTRY

Long before Laborde published his treatise on the union of art and industry, men who made a living from the creation and production of decorative objects put forward their own proposals to the government for the improvement of artistic quality in manufacturing. As soon as the Crystal Palace exhibition was announced in 1850 Jules Klagmann, a sculptor for Sèvres and a member of the Conseil Supérieur des Manufactures Nationales, requested that this council hold a separate exhibition of fine arts applied to industry. The council

56. Ibid., 808–920.
57. Ibid., 849.

approved the idea and passed along the proposal to the Ministry of Agriculture and Commerce, but the project failed to materialize at that time.[58] Klagmann, however, persisted in requesting public celebration of highly skilled artisans—industrial artists, as he called himself and his colleagues. Following the 1851 exhibition a group of artisans and manufacturers headed by Klagmann and forming the Comité Central des Artistes et des Artistes-Industriels addressed a petition to Napoleon III. In three different essays they explained their ideas and reasons for not only a special exhibition of artist-manufacturers but also for a museum of industrial fine arts and a school for the instruction of fine arts applied to industry.[59]

Klagmann wrote that while France owed its industrial greatness to its artisans—a fact the exhibition made eminently clear—society discriminated against them, chiefly by denying aspiring craftsmen the proper education in their trades. He pointed out that industrial artists were prohibited from exhibiting at fine-arts exhibitions, and that at industrial exhibitions such as the Crystal Palace they lost their artistic identity. Klagmann proposed that industrial artists should have their own exhibitions, organized along lines similar to those of industrial exhibitions, and that within exhibitions of industry there should be a special section of "fine arts applied to industry," divided into classes for architecture, sculpture, and painting. Klagmann's concern here was that men who combined their original artistic conceptions and skillful execution in useful and decorative objects should receive the appropriate recognition in the form of personal honor, pecuniary rewards, and protection against plagiarism.[60]

In addition to regular, state-supported exhibitions of their work, the committee considered education an essential means of perpetuating industrial artists' skills and the appreciation of them. The committee, like Laborde, concluded from the exhibition that France's superiority in art and taste had to be cultivated systematically in order to endure. Just as Britain and other nations responded to the exhibition by founding schools of design and museums for all types of

58. Eugène Véron, *Histoire de l'Union centrale* (Paris: Debons, [1857?]), 10.

59. Comité central des artistes et des artistes-industriels, *Placet et mémoires relatifs à la question des beaux-arts appliqués à l'industrie* (Paris: Mathias, 1852), 7–8. Signers of the petition included several painters from Sèvres, designers of shawls, a merchant, a photographer, engravers, an architect, a civil engineer, a designer of wallpaper, and many more.

60. Ibid., 10–17.

capital and consumer goods, so, too, should France establish institutions that provided models for artists and workers and that cultivated public taste. "The London exhibition should be a decisive occasion, for, if it satisfied our national self-esteem, it revealed at the same time the intense efforts that rival nations are undertaking to challenge our supremacy in matters of industrial art."[61] Charles-Ernest Clerget (b. 1812), a designer and ornamental engraver, authored the second essay on creating a museum of industrial fine arts, emphasizing its instructive role for both artists and the public by providing models of the finest work from the past and present. In the text Clerget acknowledged that others had proposed plans for such a museum earlier in the century, but he hoped that the exhibition experience would precipitate government action on the idea. For good measure he included a report on the Museum of Practical Art which opened in London in May 1852 with several acquisitions from the Crystal Palace exhibition.[62] The third essay, by a painter for the Gobelins and Beauvais tapestry works, Chabal-Dussurgey, outlined a plan for a special central school of industrial fine arts.

The committee's plans for reform differed from Laborde's in several regards, including the extent of state involvement. Though the committee initially sought government support for art education and also for museums, it was generally pessimistic about the state's willingness or ability to transform public taste. It turned instead to private initiatives like the British efforts at the time of the Crystal Palace exhibition, and later to the collaboration of political and cultural figures that created art nouveau during the Third Republic.[63] This reliance upon private as well as public support for art in manufacturing reflected the backgrounds and interests of committee members, many of whom were manufacturers in the private economic sector, or were employed by them. This collection of men did

61. Ibid., 34.

62. Ibid., 20–36. This Museum of Practical Art later became the Victoria and Albert Museum. Pevsner attributes the origin of this institution to the influence of Gottfried Semper on Henry Cole and other members of the Executive Commission of the Crystal Palace exhibition. Semper, a German contemporary of Laborde, also wrote a major treatise on improving professional education and public taste, partly in response to the 1851 exhibition. Pevsner, *Academies,* 251–55.

63. Pevsner, *Academies,* 255; Silverman, *Art Nouveau*; Archives nationales, Ministère de l'Agriculture et du Commerce, F¹² 2334, Institutions, sociétés, dons, et prix en faveur des arts et manufactures, 1830–1870.

not manifest Laborde's inflexible aesthetic, or his need for centralized and mass imposition of it. These artisans and manufacturers talked about the "unity of art" as something deriving entirely from human inspiration and cultivation, and in danger of being lost in the excessive dependence on machines to produce goods, particularly reproductions or imitations of art. Like Laborde, they did not condemn machine production out of hand; they sought the successful exploitation of machines by human beings imbued with artistic sensibility through education and exposure to fine works of art in museums and exhibitions.[64]

The committee and Laborde differed radically, however, in their conceptions of how to regard artisans and manufacturers as practitioners of art. Whereas Laborde tried to deny any distinction between artists (painters, sculptors, architects) and artisans (goldsmiths, cabinetmakers, potters, jewelry makers, etc.), asserting that they simply applied the same artistic principles to different media and for different purposes, the committee called for separate, public recognition of the accomplishments of industrial artists. Even though Laborde acknowledged some difference in the way that artists and artisans should interpret nature, given the different purposes of their finished products, he castigated the movement of industrial arts, dubbing it a "pestilence" ranking with the plagues of Egypt.[65] He considered insane the suggestion that a distinct art appropriate to industry should be taught in professional schools and displayed in special exhibitions. The self-proclaimed industrial artists, Laborde declaimed, were misguided, vain, self-pitying fools.

In fact, they were more pragmatic than Laborde, and they experienced professional insecurities that the well-placed, aristocratic civil servant did not. Some of Laborde's theories regarding the higher education of artists were implemented in a decree of 1863, but his sweeping plans to unify French taste nationwide never left the drawing board. By contrast, the committee succeeded in several of its more modest proposals, including a separate section for art industries at the Paris exhibition of 1855 and at independent exhibitions in 1861 and 1863.[66] The committee was also the precursor of the Union Centrale des Arts Décoratifs, which supported the rococo revival and

64. Véron, *Histoire*, 13–14.
65. Commission française, *Exposition* 8:411.
66. Véron, *Histoire*, 11–12, 16–18.

later the creation of art nouveau under the Third Republic.[67] Unlike Laborde, with his overriding concern to purify public taste and to impose his aesthetic on the entire French population, the committee sought primarily support for the producers of furnishings and decorative objects; and to some extent they achieved it. In any case, concern for the artistic quality of manufactured goods remained vocal and visible for the half-century following the Crystal Palace exhibition.[68]

CONCLUSION

"Art is not a luxury; it is a necessity," wrote Théodore Labourieu in 1863 as part of an appeal to Napoleon III to support exhibitions of industrial art.[69] A follower of Klagmann and the committee, Labourieu expressed a sentiment shared by many of his time. The performance of France at the Crystal Palace exhibition was a great lesson for producers and would-be policy makers, indicating that French industrial competitiveness lay in the direction of cultivating the arts. From the committee's proposals for exhibitions of applied arts to Laborde's plans for homogenization of public taste through state institutions, art reform in France was firmly linked with successful industrial development and export to foreign markets. What the exhibition revealed was that art was no longer the exclusive preoccupation of wealthy elites; along with technology, it comprised the bones and sinews of national industry which affected all French inhabitants in daily work and life—in both production and consumption.

Of all the post–Crystal Palace art reformers Léon de Laborde was the most comprehensive and the most memorable, if not the most influential. His proposals for mass education in art, the creation of a huge state manufacture, and public support for exhibitions, athletics, libraries, museums, and parks far exceeded any other reform plans in scope and ambition. Laborde proposed to use state power in order to expand state power, all in the interest of establishing what he called aesthetic stability—that is, the overwhelming of individual taste by a state-sanctioned definition of art and beauty. He wanted to see all

67. Silverman, *Art Nouveau*, 109–33.
68. Silverman, *Art Nouveau*; Levin, *Republican Art*; Williams, *Dream Worlds*.
69. Théodore Labourieu, *L'Organisation du travail artistique en France* (Paris: E. Dentu, 1863), 5.

French men, women, and children imbued with the models and principles of ancient Greek art. In this way, he anticipated that consumers would buy only "good" art and beautiful objects, which producers would gladly manufacture. Modern technology presented no terrors to Laborde in his vision of artistic renaissance, for he was certain that machines and steam power—even electricity—could be employed in the manufacture of artistic products under the condition of aesthetic stability. Presumably the rest of the world, too, would appreciate the good taste and artistic quality of French goods and flock to purchase them, abandoning the inelegant products of rival England. France's future as an industrial power, Laborde argued, depended on transforming public taste, whatever the cost in money or in private choice.

Consumption was all-important to Laborde's ideas about art and industry. Was this emphasis due to his aristocratic origins, his admiration for past epochs when wealthy elites were the only consumers, and hence the main influence upon art and industry? Or was he ahead of his time, discerning the power of the modern state to manipulate and control the public to do its bidding, in this case in matters of art and taste? Laborde's plans included both conservative and radical elements, which may explain why he is forgotten today. Laborde proposed to "purify" public taste by abandoning laissez-faire, which in Laborde's view represented consumer promiscuity and the utter dependence of manufacturing upon fickle and misguided consumer tastes. Unfortunately for Laborde, laissez-faire proved more enduring than his own ideal of aesthetic stability. Indeed, French advocates of free trade were as adamant as Laborde in deriving from the Crystal Palace exhibition support for their own position—that laissez-faire, not art reform, was the key to France's future industrial success.

Chapter Seven

Political Economists and Specialized Industrialization

After examining the machines and products from all over the world on display at the Crystal Palace exhibition, two French political economists concluded that "the United States can feed the world, England can clothe it, and France can beautify it."[1] Joseph Garnier (1813–81) and Hippolyte Dussard (1798–1876) based this assessment on their observations that the United States' major contribution to the exhibition was its agricultural products, that England excelled in the production of cheap, machine-made textiles, and that France's strength as an industrializing nation lay in the good taste and design of its manufactured consumer goods. This view was more than a reiteration of the ubiquitous praise for tastefulness in French manufacturing that the exhibition inspired; it was the foundation of a vision for industrial development in France that political economists intended to implement through an economic policy of free trade. The exhibition marked a turning point in the rhetoric of political economy, as adherents emphasized small-scale, hand production as a sound basis for French economic growth, both to compete in world markets and to encourage social stability at home. Like Léon de Laborde and the artisans studied in chapter 6, political economists also devised a plan for the perpetuation of good taste and quality in French manufacturing as a result of the nation's performance at the Crystal Palace. But their plan reflected a decidedly bourgeois perspective, and it must be seen as part of the process whereby this class became dominant in French economics, politics, and society over the course of the nineteenth century.

1. *Journal des économistes* 30 (1851): 133.

Few scholars have credited any group, least of all political economists, with a conception of small-scale, parcelized, hand manufacturing as a viable means of industrialization in France.[2] With their
Saint-Simonian love of technology, their belief in individualism and
careers open to talent, and their unswerving support of free-market
capitalism, political economists at first glance appear to be unlikely
connoisseurs of art and taste or advocates of the manufacturing
methods that promoted them.[3] Yet upon closer examination their
appreciation for France's unique competencies in comparison with
other industrializing countries was in fact consistent with their long-

---

2. Economic historians who treat the persistence of handicraft manufacture in
French industry as a purely pragmatic response of French producers to domestic
market conditions include Alain Faure, "Petit Atelier et modernisme
économique: La Production en miettes au XIXe siècle," *Histoire, économie et société*
4 (1986): 531–57; Patrick O'Brien and Caglar Keyder, *Economic Growth in Britain
and France, 1780–1914* (London: George Allen and Unwin, 1978); Richard Roehl,
"French Industrialization: A Reconsideration," *Explorations in Economic History* 13
(1976): 233–81; Maurice Lévy-Leboyer, "Les Processus d'industrialisation: Le
Cas de l'Angleterre et de la France," *Revue historique* 239 (1968): 281–98; Rondo
E. Cameron, "Economic Growth and Stagnation in France, 1815–1914," in *The
Experience of Economic Growth: Case Studies in Economic History,* ed. Barry E.
Supple (New York: Random House, 1963), 328–39. Scholars who see some
contemporary justification for this pattern include Alain Cottereau, "The Dis-
tinctiveness of Working-Class Cultures in France, 1848–1900," in *Working-Class
Formation: Nineteenth-Century Patterns in Western Europe and the United States,* ed.
Ira Katznelson and Aristide R. Zolberg (Princeton: Princeton University Press,
1986), 111–54; Louis Bergeron, "French Industrialization in the Nineteenth
Century: An Attempt to Define a National Way," *Proceedings of the Annual
Meeting of the Western Society for French History* 12 (1984): 154–63; Francis Demier,
"Adolphe Blanqui, 1798–1854: Un Économiste libéral face à la Révolution
industrielle" (Ph.D. diss., University of Paris, X–Nanterre, 1979).
   3. Joan W. Scott, "Statistical Representations of Work: The Politics of the
Chamber of Commerce's *Statistique de l'industrie à Paris, 1847–48,*" in *Work in
France: Representations, Meaning, Organization, and Practice,* ed. Steven Laurence
Kaplan and Cynthia J. Koepp (Ithaca, N.Y.: Cornell University Press, 1986),
335–63; Yves Breton, "Les Économistes, le pouvoir politique et l'ordre social en
France entre 1830 et 1851," *Histoire, économie et société* 4 (1985): 233–52; William
M. Reddy, *The Rise of Market Culture: The Textile Trade and French Society,
1750–1900* (New York: Cambridge University Press, 1984), chap. 6; William
Coleman, *Death Is a Social Disease: Public Health and Political Economy in Early
Industrial France* (Madison: University of Wisconsin Press, 1982); Dennis Sher-
man, "The Meaning of Economic Liberalism in Mid-Nineteenth-Century
France," *History of Political Economy* 6, no. 2 (Summer 1974): 171–99; Michel
Lutfalla, "Aux origines du libéralisme économique en France, le *Journal des
économistes:* Analyse du contenu de la première série, 1841–1853," *Revue d'histoire
économique et sociale* 50 (1972): 494–517.

term commitment to free trade and economic growth. Moreover, the revolution of 1848 reinforced political economists' emphasis on small-scale and hand production as a means of avoiding future social unrest.

It is impossible to understand political economists' response to the Crystal Palace exhibition, and in particular their support for hand and handicraft manufacturing in France, without addressing their views of class structure and social relations. This chapter focuses on the social relations of production, and especially the class conflict that industrial development aggravated; for political economists hoped to resolve such conflicts through specialized industrialization and free trade. For some of these men, the Crystal Palace exhibition of 1851 was an antidote to the revolution of 1848 — a peaceful, bourgeois celebration of industrial capitalism instead of a violent, working-class rebellion against it. They sought to exploit France's success at the exhibition for their goals of economic reform and social order under bourgeois leadership. A look at the preexhibition free-trade campaign of political economists, and at their social origins and career patterns, will help make this clear.

POLITICAL ECONOMISTS AND FREE TRADE

Modern political economy, beginning with Adam Smith (1723–90) and disseminated in France by Jean-Baptiste Say (1767–1832), was primarily concerned with the creation of wealth. Fundamental to political economy was the principle of laissez-faire, that only in a completely unregulated market could wages and prices reach levels that would guarantee the greatest possible benefits to employers, workers, and ultimately the state. Adam Smith and, to a lesser degree, Say laid the theoretical foundations of political economy at a time when industrial capitalism was just beginning; subsequently Say's followers in France tried to influence French economic policy to conform with economic principles. Confronting the dismal reality of unemployment, urban slums, poverty, and disease affecting the working class in France under the July Monarchy (1830–48), Say's successors explained these problems as the result of either the moral failure of the poor or the misguided policies of the French government. Education was one way of combating the ignorance and degeneracy that some political economists blamed for unemployment and poverty. But men influenced by political economy concentrated more on economic than on social solutions to the problems of early industrialization. In the 1840s they launched a campaign to abolish protectionism and establish free trade, expressing their views in the *Journal des économistes* and through the Association pour le Libre Échange. Additionally, professors of political economy spread the word in their courses, in other periodicals, and in the Académie des Sciences Morales et Politiques de l'Institut Français. They argued that protectionism artificially raised the prices of all goods and that it smothered competitiveness, innovation, and growth among manu-facturers. Under protectionism, so they claimed, capital formation was inhibited, industry did not grow sufficiently to employ all work-ers, and consumers could not buy all that they wanted or needed.[4] In theory, then, protectionism violated economic laws; and political economists maintained that only free trade would restore the natural operation of these laws to ensure maximum productivity and pros-perity.

What made these men turn from theory to politics, and what did the free-trade movement represent to them? Brief biographies of some of the more prominent free traders, including two of the most influential political economists of the middle of the nineteenth cen-tury in France, help explain this development. All of these men were political activists as youths, opposing the policies of the Restoration government (1814–30) that threatened the hard-won gains of the French Revolution. Almost all of them came from modest, middle-class backgrounds and rose to positions of prominence in academics and politics through hard work, education, and ambition. They shared the values of individualism, effort, and social responsibility. Free trade, for these men, was a reform that would increase national wealth, raise the standard of living, and maintain a social and eco-nomic structure based on bourgeois dominance and private property. The class backgrounds, career patterns, and political experiences of the political economists suggest a complex of motives behind the free-trade movement and explain the alacrity with which these men formulated the concept of specialized industrialization, as much to satisfy political goals as to implement economic principles.

The main figures in this study are Adolphe Blanqui (1798–1854) and Michel Chevalier (1806–79), two of the better-known political

4. Lutfalla, "Aux origines."

economists, who represented the Académie des Sciences Morales et Politiques at the Crystal Palace exhibition. Adolphe Blanqui, elder brother of the insurrectionist Auguste Blanqui, made his way to Paris as a young man after his father lost the position of subprefect in the Alpes-Maritimes with the Restoration. Working diligently at medical studies and teaching in secondary schools, Blanqui met the famous economist Jean-Baptiste Say and became his protégé. Through Say's influence Blanqui became a professor at, and later director of, the École de Commerce, and in 1833 he succeeded Say in the chair of political economy at the Conservatoire des Arts et Métiers. On the basis of his writings, Blanqui was elected to the Académie des Sciences Morales et Politiques in 1838. As a consistent advocate of free trade in the classroom and in print, Blanqui became the representative of Bordeaux to the Assemblée Nationale in 1847; but the February Revolution of 1848 ended his short career in elected office.[5]

Blanqui's junior colleague at the Académie des Sciences Morales et Politiques as of 1851, Michel Chevalier, was the son of a Limoges tax collector. Chevalier began to study at the École Polytechnique in Paris when he was eighteen, and two years later went on to the École des Mines. After a period of serious involvement with Saint-Simonism, Chevalier accepted an assignment to study railroads in North America, followed by several positions on advisory councils. In 1840 he took the chair of political economy at the Collège de France. Because of his ardent support of free trade, and especially because of his criticism of socialism, Chevalier temporarily lost his position at the Collège de France in 1848. He soon regained the chair, only to move on to a position on the Conseil d'État, the better to influence Louis-Napoleon Bonaparte's government toward a free-trade economic policy.[6]

The other political economist to visit the Crystal Palace in an official capacity was Louis Wolowski (1810–76), who was a member of the French jury at the exhibition. Wolowski was born in Warsaw,

5. Demier, *Adolphe Blanqui;* Alan Spitzer, "The *Mal du Siècle* of Adolphe Blanqui" (Paper presented at the Annual Meeting of the Society for French Historical Studies, Los Angeles, 22–23 March 1985).

6. Jean Walch, *Michel Chevalier économiste saint-simonien, 1806–1879* (Paris: J. Vrin, 1975); J.-B. Duroselle, "Michel Chevalier saint-simonien," *Revue historique* 215 (1956): 233–66; Arthur Louis Dunham, *The Anglo-French Treaty of Commerce of 1860 and the Progress of the Industrial Revolution in France* (Ann Arbor: University of Michigan Press, 1930).

the son of a respected lawyer and member of the Polish Diet. He studied in Paris as a youth, but returned to Poland in 1830 to fight with the Republicans in the revolution there. Escaping back to France, Wolowski learned that his possessions had been confiscated and that he had been condemned to death. He became a French citizen after finishing his law degree in Paris, and eventually became an expert on law and legal history. In 1839 Wolowski took the chair of industrial legislation at the Conservatoire des Arts et Métiers; he succeeded Blanqui in the combined chair of political economy and industrial legislation in 1860. A committed republican, Wolowski was elected deputy from the Seine in 1848, but true to the tenets of political economy he consistently opposed all measures smacking of socialism.[7]

Other supporters of the Crystal Palace exhibition, free trade, and political economy during the Second Republic were Joseph Garnier, Léon Faucher (1803–54), and Hippolyte Dussard. Garnier's career closely resembled that of Blanqui's. The son of a farmer, Garnier went to Paris as a young man and became Blanqui's secretary as well as a teacher at the École de Commerce. In 1843 Garnier became director of the *Journal des économistes,* which he edited, with one interruption, for the rest of his life. In 1846 Garnier became the first appointee to the chair of political economy at the École des Ponts et Chaussées, and later he became a member of the Académie des Sciences Morales et Politiques and senator from the Alpes-Maritimes.[8]

Faucher, too, came from modest, even straightened, provincial circumstances and went to Paris to study and teach. The 1830 revolution interrupted his education, and for several years thereafter Faucher made a living as a journalist. In the 1840s he was a leader in the Association pour le Libre Échange (and he married Wolowski's sister). Faucher turned to politics in 1847, rising to the position of minister of the interior during Bonaparte's presidency before leaving politics in advance of the coup of 2 December 1851.[9]

What is known of Hippolyte Dussard's background is sketchier than those of his peers. An outspoken opponent of the 1830 ordi-

7. Michel Lutfalla, "Louis Wolowski ou le libéralisme positif," *Revue d'histoire économique et sociale* 54, no. 2 (1976): 169–84; *Journal des économistes,* 3d ser., 44 (1876): 321–45.

8. *Journal des économistes,* 4th ser., 16 (1881): 5–13.

9. *La Grande Encyclopédie* (Paris, 1893) 17:38–40.

nances censoring the press, Dussard later joined political economists in editing the *Journal des économistes* from 1843 to 1845. In 1848 he was appointed prefect of the Seine-Inférieure, and thereafter served the Second Republic on the Conseil d'État and as investigator of charity and public assistance in England.[10]

With the exception of Wolowski, and possibly of Dussard, these professional and amateur political economists were all self-made men (and even Wolowski had to overcome penury after the 1830 revolution). They were beneficiaries of the 1789 revolution, which opened careers to bourgeois men of talent, and were committed to a social and political order that maintained these open channels of upward mobility and protected private property. With maturity and professional success under the July Monarchy, these men abandoned their youthful politics of revolutionary opposition and focused on economic reform—specifically free trade—as the means to ensure bourgeois access to and domination of economic and political power. Moreover, they believed free trade to be the antidote to working-class misery and therefore rebellion.

Political economists strongly supported technological progress and economic growth in France, while they deplored the growing number of unemployed, slum-dwelling workers around new industrial cities. Those who tended toward moral explanations of poverty more or less accepted it as an inevitable if unfortunate consequence of industrialization. Léon Faucher, for one, claimed in typically blunt fashion: "I recognize that mechanical power, in developing the resources of industry, leads to violent rendings in the social order. Steam [power], like the cannon, blows holes in the masses."[11] But if the state could not totally avert the casualties of mechanization, several political economists believed it could at least alleviate the workers' condition through economic reform to stimulate demand and employment. They thought that free trade, what Louis Wolowski called "this positive protection," would accomplish that goal and benefit both workers and industrialists.[12]

Of greater concern to political economists, however, was the socialist challenge to the system of private property and free enterprise. In 1845 a committee of political economists reported: "Perhaps

10. *Journal des économistes*, 3d ser., 41 (1876): 302.

11. Joseph Garnier, ed., *Le Droit au travail à l'Assemblée nationale* (Paris: Guillaumin, 1848), 331.

12. Ibid., 364.

at no other time in history has the teaching of the laws of political economy been more useful, more necessary, more indispensable. . . . The current foundations of society are being attacked everywhere . . . by fallacies that tend toward nothing less than the weakening of respect for property and [that] put society in danger."[13] Countering the socialist demand for a radical transformation of economic and social relations to end poverty, unemployment, and wage labor, political economists offered moderate reform to improve the workings of laissez-faire capitalism for the benefit of all society.

With the revolution of 1848 political economists could only repeat their criticisms of socialism and their advocacy of free trade, but to no avail. And during the Second Republic, after the repression of the June Days, they increasingly confronted not just radical alternatives to their proposals but also conservative opposition to any change that threatened the restored order. Barraged from both left and right, proponents of free trade found in the Crystal Palace exhibition of 1851 new evidence supporting their cause as well as the outlines of an industrial future for France that would, in principle, satisfy workers, protectionists, and political economists with the continuation of handicraft and hand manufacturing.

THE EXHIBITION AND SPECIALIZED INDUSTRIALIZATION

Political economists welcomed the first international exhibition of industry from its inception, for it perfectly conformed to their ideas of free trade. They were particularly attentive to the comparison of French and British exhibits, considering Britain to be France's chief rival for industrial supremacy; and they were not disappointed in the respectable number of prize medals and honorable mentions awarded to French manufacturers.[14] The supporters of free trade, like other

13. Quoted in Breton, "Économistes," 237.
14. The total number of awards for France, including prize medals and honorable mentions, was 1,050, compared to 2,089 for Britain, 482 for the *Zollverein*, and 236 for Austria. A total of 5,186 prizes were distributed; there were 18,000 exhibitors from all countries. French officials calculated that for every 100 French exhibitors France obtained 66 awards, while for every 100 British exhibitors Britain obtained 25 awards. Ministère de l'Intérieur de l'Agriculture et du Commerce, *Annales du commerce extérieur: Faits commerciaux*, No. 20, *Exposition universelle de Londres en 1851* (Paris: Imprimerie impériale, 1853), 24.

observers, agreed that the significant feature of the French exhibits
was their fine quality. In Blanqui's words:

> The main result of the exhibition for the French is the universal,
> absolute, uncontested recognition of their superiority in matters of art
> and taste. In brocaded or printed fabrics, in cabinetwork, in gold
> work, in the manufacture of bronzes, of wallpaper, of porcelain—they
> have no rivals at all.[15]

The arrangement of exhibits by country facilitated the comparison of
industry in England and France, and from this comparison political
economists drew the following conclusions.[16] The English excelled in
the use of mechanical means, in the investment of large amounts of
capital in industry, and in the production of mass quantities of
low-priced goods. By contrast, French superiority lay in the good
taste and skill of handicraft and hand workers, and in the fine quality
of expensive luxury items.[17] Chevalier summarized the comparison
thus: "The Frenchman engages in industry as an artist, the English-
man as a merchant."[18]

A clear indication to Chevalier and his colleagues of the artistic
bent of French manufacturers was the small-scale and hand production methods of most industrial establishments.[19] As the *Dictionnaire
de l'économie politique* explained in its definition of "manufacturing
industry," a true idea of its importance

> resides . . . in the infinite number of second- or third-order work-
> shops; in those little manufacturers, artisans, craftsmen of all types;
> shops that, of little consequence when taken in isolation, prevail over

15. Adolphe Blanqui, *Lettres sur l'Exposition universelle de Londres* (Paris: Capelle, 1851), 107.

16. Dussard noted that the arrangement of exhibits by country facilitated national comparisons. He suggested that organizing products by category would have been more scientific than by country, but "less striking to the imagination" and "not . . . as grand a lesson in history and political economy." *Journal des économistes* 30 (1851): 128–29.

17. Michel Chevalier, *L'Exposition universelle de Londres* (Paris: Mathias, 1851), 34; Blanqui, *Lettres,* 178–84, 200–207; *Journal des économistes* 30 (1851): 133.

18. Chevalier, *Exposition,* 34.

19. T. J. Markovitch, "Le Revenu industriel et artisanal sous la Monarchie de Juillet et le Second Empire," *Économies et sociétés,* série AF-8 (April 1967): 85–87.

the others by their number, so that as a whole they present a mass of work far superior to what is done in large manufactures.[20]

The exhibition showed that handicraft or hand manufacturing on a small scale was responsible for the art and taste in French manufactured goods; indeed, political economists suggested that this type of production reflected the creative and artistic temperament of French workers, in contrast to the more passive and unimaginative English workers who labored to the rhythm of machines.[21] These national characteristics of workers, seen as embodied in the exhibits at the Crystal Palace, were integral to the political economists' conception of future industrial development in general, but especially in France.

Political economists believed that natural and human resources were divided among different countries and different parts of the world, requiring complete freedom of trade for the earth's inhabitants to enjoy all the riches nature provided. "Providence has spread with unparalleled liberality over all the surface of the earth everything that is necessary for the subsistence and the ease of man. . . . Our task consists of exchanging, from pole to pole, the liberalities of nature."[22] In this schema, each product and each type of production had its niche; the political economists' objective was not to alter this distribution through uniform industrial progress, but to make all varieties of goods available to all people through free trade. Thus they believed that each country should develop its own special skills and products in anticipation of international freedom of exchange; for France this meant encouraging the production of artistic and tasteful consumer goods. "The true prosperity of our country . . . rests on the progressive development of its natural industries, that is . . . the arts on which skillfulness of hand and purity of taste can exert their influence."[23]

What about the development of technology and large-scale, mechanized production? On this point the political economists were vague. In general they favored technological advancement and indus-

20. Coquelin and Guillaumin, eds., *Dictionnaire de l'économie politique* (Paris: Guillaumin, 1853) 1:932.
21. Blanqui, *Lettres,* 200–207.
22. Coquelin and Guillaumin, *Dictionnaire* 1:749.
23. Blanqui, *Lettres,* 283.

trial progress and believed that free trade would stimulate such changes through competition:

> Competition . . . is the industrial side of liberty. . . . It is the rivalry that sustains invention and provokes the best use of capital, improvements, and progress of all kind, in increasing the amount of all products and services and in thus permitting the largest number of men to procure for themselves and their families growing quantities of material, intellectual, and moral satisfactions.[24]

But Garnier's emphasis here is less on the *means* of increasing the production of consumer goods than on the *end* of more goods and services for more people. Political economists never suggested, for instance, that French industry in general should follow the English model, investing in machinery in order to increase output and thus lower the price of goods. To be sure, lower prices were a goal they thought French producers should adopt, but they argued that free trade alone would accomplish this by bringing down the cost of raw materials. In addition, free trade would reduce the prices consumers had to pay for foodstuffs, thus releasing part of their income for the purchase of manufactured goods.[25] Political economists were not concerned about the influx of cheap British products into France. They believed that domestic producers of textiles, metal products, and machines would improve to become competitive with the British, or that domestic producers would switch to some other form of manufacturing—such as the production of luxury goods—where Britain was less competitive. In either case the outcome would be favorable to both producers and consumers in France.

From Adam Smith onward, political economists were preoccupied with the interests of consumers rather than those of producers. The goal of political economy was to maximize the production of wealth in order to improve the standard of living in all of society, but neither Smith nor the French political economists assumed that this required the expansion and concentration of industries to the exclusion of small-scale production. Indeed, from the political economists' perspective, the French pattern of industrialization, based on the small-

24. *Journal des économistes* 29 (1851): 180.
25. Chevalier, *Exposition*, 5, 27–28; Blanqui, *Lettres*, 64–65, 97, 103.

scale manufacture of tasteful and artistic goods, was the only one that could satisfy French consumers.[26] For example, when Blanqui compared English and French shawls at the exhibition, he asserted that the taste of French women for articles of beauty and art prevented them from buying inferior English goods: "I beg pardon of our English neighbors; all these printed shawls, really shoddy shawls [*chales de pacotille*] . . . would never be worn in Paris even by chambermaids of a good household."[27] Wolowski, too, asserted a certain discrimination on the part of French consumers when he noted, in his report on the wallpaper industry, that new mechanized and steam-powered techniques of producing cheap and simply-patterned wallpaper would not succeed in France because consumers preferred more intricate designs, which could only be produced by hand printing methods.[28] Thus political economists saw no imperative for French manufacturers to expand and mechanize industry for the production of cheap, shoddy goods. Presumably, the working poor, if they wanted such articles, could buy them from England under a free-trade economic policy. But by and large political economists considered workers as primarily producers rather than consumers; though they argued that workers, too, would benefit from lower food prices with free trade. However, when speaking of consumers, what they primarily had in mind was the bourgeoisie.

Regarding the future industrial development of France, their main prescription was the continuation of what France did best. As Garnier concluded from the exhibition:

> A lesson can be drawn, especially for France, from the demonstration in Hyde Park; that she is, on pain of decline, summoned by her future and the interest of her glory and wealth, to put an end to public agitations and then to modify her commercial and financial regime, her economic legislation, her public education, so that the genius of her workers can continue to be both the rival and the model of other nations.[29]

26. Whitney Walton, "'To Triumph before Feminine Taste': Bourgeois Women's Consumption and Hand Methods of Production in Mid-Nineteenth-Century Paris," *Business History Review* 60 (Winter 1986): 541–63.

27. Blanqui, *Lettres,* 135.

28. Commission française sur l'Industrie des Nations, *Exposition universelle de 1851: Travaux de la Commission française sur l'Industrie des Nations* (Paris: Imprimerie impériale, 1855) 7:7–14.

29. *Journal des économistes* 30 (1851): 133–34.

Garnier implied here that economic and educational reform would allow French handicraft and hand workers to continue producing quality goods and thus maintain their superiority over other countries. But Garnier, like other political economists, was also concerned by the potential of French workers for rebellion, and saw this as a further justification for specialized industrialization.

SPECIALIZED INDUSTRIALIZATION AND WORKERS

Although the exhibition revealed the talent of French handicraft workers for producing award-winning articles, political economists remembered that workers in consumer goods industries had led the revolution of 1848 and had greatly impeded economic recovery after the depression years of 1846–47. Furthermore, many artisans and skilled workers espoused socialist ideas and called for state guarantees of employment and support of cooperative production. Political economists often had difficulty reconciling these two aspects of the French working class—their skill and their subversiveness—because they did not fully comprehend the effect of industrialization on handicraft and skilled workers. The effect of mechanization on unskilled workers in new industrial centers was obvious to political economists; subjected to the implacable rhythm of machines, workers lost control over their own labor and over the profits it generated, which were appropriated by the capitalist owners of the machines.[30] This situation would further deteriorate when market slumps caused mass unemployment; mass unemployment, in turn, would bring down wages, and working-class families would sink into debauchery, degradation, and rebelliousness.[31] Political economists regarded this situation as necessary and temporary, and one that would eventually improve with general economic growth. But the causes of discontent among artisans and handicraft workers, leading them to socialism, were less clear. Economic theorists failed to discern the decline in status that accompanied the manufacturers' capitalist practices of division of labor, underemployment, and lower wages. Faucher could come up with only the old moralistic explanation

30. Adolphe Blanqui, *De la concurrence et du principe d'association* (Paris: Panckoucke, 1846), 5.

31. Adolphe Blanqui, *Des classes ouvrières en France pendant l'année 1848* (Paris: Firmin Didot frères, 1849), 105–6.

when he tried to analyze the spread of revolutionary socialism in the working class during 1848:

> [Socialism] began in the luxury industries, where the workers, even though earning exceptional wages, found themselves confronting frequent unemployment; where the irregularity of the labor force gave way to evil passions and laziness. It spread later and by degrees to the powerful wool and cotton industries.[32]

The repression of the June Days in 1848 had, as Faucher said, vanquished socialism in the streets; but he also asserted that political economists must combat it intellectually; and the Crystal Palace exhibition contributed significantly to that effort. Political economists could hope to reach only a small proportion of workers through their teaching and publications, but the Crystal Palace exhibition brought universal acclaim to French handicraft manufacturers and workers. It was thought that this recognition would be crucial not only in persuading workers of the good intentions of political economists and the benefits of free-market capitalism but also in convincing the French bourgeoisie as a class of the wisdom of free trade and specialized industrialization as means of preventing popular revolution. The exhibition clearly demonstrated the essential role of the workers in French handicraft production: "It is in studying the craft process [*procédés des arts*] and the actual role of the workers in it, that one adequately appreciates their contribution to these admirable works of which our country is so justly proud."[33] This recognition, Dussard thought, marked the beginning of a new era in which free trade would amend the negative effects of protectionism and "the worker will soon be . . . in full possession of his labor, in full security for his future."[34] But how did political economists transform the image of the workers from violent revolutionaries to satisfied artists in the eyes of the French bourgeoisie? And why was this so important to their arguments for free trade and specialized industrialization?

In their writings political economists minimized the disruptive and threatening features of artisans and handicraft workers by denying

32. Garnier, *Droit,* 330.
33. Blanqui, *Lettres,* 197.
34. *Journal des économistes* 29 (1851): 39.

their working-class status.[35] Michel Chevalier, for instance, avoided the word *ouvrier,* with its connotations of manual labor—in favor of the broader term *travailleur,* which for Chevalier included industrialists, intellectuals, artists, and bureaucrats, as well as manual laborers.[36] According to Chevalier and other political economists, anyone who contributed skills, intelligence, or labor to the production of national wealth was a worker in the broadest sense of the word, a notion deriving from the Saint-Simonian distinction between producers and parasites (the latter exemplified by the idle aristocracy of the Old Regime). Political economists further tried to identify working-class interests with those of the bourgeoisie by defining workers as potential entrepreneurs. In the *Dictionnaire de l'économie politique,* Garnier defined *ouvriers* as those who labor under the direction of an entrepreneur who provides raw materials and the means of production; in return for this labor the workers earn a wage. However, Garnier went on to say that many workers owned their own tools—a form of capital—and in certain industries provided some raw materials; and this made them "capitalists and entrepreneurs."[37] Garnier was certainly correct in noting that a large proportion of workers owned their own tools and/or had small shops from which they sold goods directly to the public, but to equate such petty artisan entrepreneurs with capitalists was to blur an important distinction. Artisan entrepreneurs still labored alongside their employees (if they had any), brought more manual skill than capital to the enterprise, and rarely made big profits from their businesses; the same did not apply to the owners of larger, mechanized industries. Political economists, in trying to minimize class differences, seem to have been trying to ignore class hostility. It was in their interest (and in the interest of free trade) to suggest to their fellow bourgeois as well as to workers that they all would benefit equally from the prosperity that would result from economic reform.

The exhibition allowed political economists to suggest also that workers and bourgeois shared the same interest in a pattern of indus-

35. Scott, "Statistical Representations," 351.

36. Michel Chevalier, *Lettres sur l'organisation du travail* (Paris: Capelle, 1848), 1.

37. Artisans, according to this definition, were not *ouvriers* at all, because they worked on and with materials they procured for themselves and also sold the finished product themselves. Coquelin and Guillaumin, *Dictionnaire* 1:301–2.

trialization that perpetuated handicraft and hand manufacture of quality consumer goods as well as developing new technology and other capitalist strategies in certain industries. The political economists' lesson from the exhibition was that specialized industrialization under a free-trade economic policy would guarantee French success in domestic and foreign markets. In addition, they argued that this type of development would solve working-class problems and so remove the causes of labor unrest and rampant socialism. Blanqui explained the merits of specialized industrialization as follows:

> The Universal Exhibition . . . provided us with new arguments in support of commercial freedom, along with a great political advertisement. It has been demonstrated . . . that the sum of values created by small industries is larger than that created by large ones, and that small industries require less capital, employ more hands, develop more intelligence, and secure more well-being, with fewer social complications, than the processes of manufacturing organized under the sway of machines and division of labor pushed to an extreme.[38]

Blanqui argued that free trade would reduce the cost of raw materials for producers and expand foreign markets for manufactured goods. In this way the majority of French workers would have employment and job satisfaction, and France would uphold its reputation for quality consumer products. The exhibition showed the way to the solution of all France's industrial problems, as far as political economists were concerned; free trade and the perpetuation of small-scale manufacturing would allow all French producers to heed Blanqui's plea to Saint-Antoine cabinetmakers after commenting on their fine show at the Crystal Palace: "O inimitable workers, if only you would make more furniture and fewer revolutions!"[39]

Political economists used the results of the Crystal Palace exhibition to support their campaign for free trade and to incorporate into this effort a vision of France's future industrial development that rested on actual conditions and on solutions to actual problems in French industry. But what about the owners of large-scale, mechanized industries in France? What about French farmers? Here, too, the

38. Blanqui, *Lettres,* 241.
39. Ibid., 50.

political economists used the exhibition to explain the benefits of free trade.

THE EXHIBITION AND PROTECTIONISM

Many of political economists' responses to the exhibition directly addressed the arguments and concerns of protectionists. Industrialists—especially in cotton textiles, metallurgy, and machine building—feared that an influx of cheaper British goods into France would ruin them. French farmers, too, supported protectionism, but it was primarily large-scale industrialists who led the campaign against free trade, and occasionally against political economists themselves. Throughout the July Monarchy the writings and teachings of political economists had attacked protectionism in theory, but with the exhibition their arguments became more practical and concrete. In their accounts of the exhibition, political economists explained the merits of free trade in the following ways: (1) the example of British economic success since the abolition of the Corn Laws in 1846 provided an important model for France; (2) the level of British technology and the modest quality of British goods at the exhibition showed that French manufacturers had nothing to fear from British competition; (3) protectionism might have served a useful purpose in earlier times and under different circumstances, but it was no longer appropriate in an age of civil equality and equality of opportunity. Each of these arguments will be examined in turn.

Britain's adoption of free trade in 1846 and its subsequent hosting of the exhibition, where Britain dazzled the world with its many machines and goods, provided a model of the benefits of free trade, according to political economists. French officials had proposed and, under pressure from protectionists, had rejected the idea of an international exhibition of industry in France in 1849; English industrialists then took the initiative, and received the credit for one of the most spectacular events of the nineteenth century. But the exhibition was only one, though by far the most impressive, consequence of Britain's adoption of free trade. Blanqui took a trip to the English countryside during the exhibition, with the intention of discovering how free trade affected English agriculture. His conversation with one farmer provided the evidence he sought; this farmer had begun raising pigs when the price of wheat fell after 1846. Blanqui was

delighted to conclude from this that since English farmers successfully adapted to free trade, French farmers could do the same.[40]

Political economists also claimed that Britain had avoided revolution because of free trade. They argued that French protectionists'
resistance to economic reform had aggravated working-class problems to the point where workers' anger exploded into the revolution
of 1848. By contrast, Britain provided a model of how to address and
solve the same problems, as Hippolyte Dussard explained in a remarkably revealing passage:

> Pressured by the shortage of 1847, aware of general hardship, [Britain]
> did not stupidly think of accusing the outcry of the press, the agitation
> of Chartism, etc. She heard their cries; but as a good mother, who
> does not discipline the child who cries, but looks for the source of the
> problem that is causing the crying, she went to the bottom of things,
> and . . . she proclaimed freedom of trade of foodstuffs, she imple-
> mented the principle of free trade.[41]

Dussard's chronology is confused; the protectionist Corn Laws
were abolished in 1846, *before* "the shortage of 1847." But it is clear
from this quotation that political economists equated workers with
helpless, frustrated children who are incapable of caring for themselves. In the metaphor, the government, enlightened by political
economy, assumes the role of a wise and attentive mother who can do
for her children what they cannot do for themselves. But political
economists alone—either in Britain or in France—could not pass the
reform legislation that would end social unrest. Only a more or less
united bourgeoisie could do that, through their access to and control
over so many resources useful to the country. "[The bourgeoisie]
represents not only most of the property, but also the intelligence, of
the nation, the understanding of business and . . . most of the private
virtues. Precisely because of this superiority, . . . it is up to it, much
more than to the workers, to do what is necessary to save France!"[42]
According to this view, political economists and industrialists together (along with other middle-class professionals and merchants)
constituted the legitimate ruling class of France, the only group

40. Ibid., 95–102.
41. *Journal des économistes* 29 (1851): 34.
42. Chevalier, *Lettres,* 322.

capable of learning from the British example. "England has reaped what she has sown," Chevalier asserted after the exhibition. "She has traversed the recent revolutionary period without receiving a blow. She offers to nations a model to follow."[43]

While political economists used the British model to argue that free trade served the collective bourgeois interest in social order and the protection of private property, they also culled from the British exhibits in the Crystal Palace evidence that French industrialists could withstand competition on a free market. In his published report from the exhibition Michel Chevalier inventoried a long list of industries represented by machines or goods in the Crystal Palace, and he concluded that British technology was no better than that of France (and that French products, though higher in price, were usually of higher quality than the equivalent British products). "The Crystal Palace is a good place for verifying the similarity, the brotherhood, the equality of industry among the main peoples of Western civilization."[44] In cotton textiles, for example, an industry that England clearly dominated, Chevalier indicated how well Alsatian producers of calicoes did compared to England. From this he suggested that all French cotton textile producers were, in terms of technology and the quality of goods, potentially the equals, if not the superiors, of English cotton textile manufacturers; and he dismissed the differences in the respective prices of goods as a mere matter of calculation.[45] The persuasiveness of Chevalier's arguments is debatable, but the import is obvious: he was trying to conciliate protectionists and allay their fears with specific examples of French success at the exhibition, both in small-scale and, especially, in large-scale industries.

Finally, political economists invoked history in their arguments against protectionism. Contrary to their earlier condemnation of protectionism as universally and at all times harmful, they now conceded that under the circumstances of war and blockade during the Napoleonic Empire, protectionism had served a useful political if not economic purpose.[46] But times had changed since then; Chevalier contended that in the context of increasing political liberty, restrictions on economic freedom were anachronistic: "Principally as a

43. Chevalier, *Exposition,* 41.
44. Ibid., 17.
45. Ibid., 24–25.
46. Demier, *Adolphe Blanqui,* 1127–33.

result of prejudices to which a desperate war gave birth, the mode of protection adopted for industry . . . is presently, at least in France, the same that flourished under the Old Regime, which naturally reflects illiberal genius and the spirit of monopoly."[47] Chevalier was suggesting here that protectionists betrayed the historical mission of the bourgeoisie by reverting to Old Regime monopoly after having successfully abolished all such privileges in 1789 and 1791. Political economists maintained that the protectionists' restriction of the free market was just as bad as the socialists' demands for state regulation of the economy; protectionists, and specifically industrialists, "presume to have a right to the tariff just as the workers [presume to have] a right to work."[48] The exhibition precipitated these particular arguments because the protectionists' refusal to exhibit was an economic and political embarrassment to political economists. If protectionism really was effective in allowing French industries to develop, Dussard reasoned, then industrialists owed it to the government to exhibit and prove the validity of the principle of protectionism.[49] Political economists interpreted industrialists' refusal to participate in the exhibition as a sign of weakness that reflected badly on France's stature as a leading industrializing nation. It also revealed a serious internecine conflict within the ruling bourgeoisie that political economists feared might weaken bourgeois domination. Winning over industrialists to the cause of free trade and specialized industrialization was the political economists' self-appointed mission; the exhibition was used as a storehouse of concrete examples and practical arguments in support of that mission.

CONCLUSION

The exhibition provided political economists with a vision of France's industrial future, based on hand manufacturing, that was basically absent from their earlier economic theories of growth and development.[50] They thought that this vision would naturally materialize

47. Michel Chevalier, *Examen du système commercial connu sous le nom du système protecteur,* 2d ed. (Paris: Guillaumin, 1852), 4.
48. Coquelin and Guillaumin, *Dictionnaire* 1:748.
49. *Journal des économistes* 29 (1851): 38.
50. Blanqui recognized the merits of small-scale industrial production for France in 1838. Adolphe Blanqui, *Cours d'économie industrielle, 1838–1839* (Paris:

with the abolition of protectionism in France. They waited for almost a decade after the exhibition to see their dream of free trade and specialized industrialization realized, however. During this period a new group of manufacturers—those involved in the railroad-building boom—joined the ranks of free traders because they wanted cheaper raw materials. The producers of luxury and export goods favored free trade, with its promise of expanded foreign markets. In addition, after 1852 Emperor Napoleon III not only had an economic and political interest in implementing free trade for France, he had also the power to do so without legislative approval.[51] After Michel Chevalier had laid the diplomatic and political groundwork for it, Napolean III signed a treaty with Britain in 1860 that inaugurated thirty years of free trade for France.[52]

Did the 1860 free-trade agreement fulfill political economists' expectations? There is no easy answer. Paul Bairoch offers a qualified no, pointing out that overall French economic growth slowed during the free-trade period, thus dashing liberal economists' hopes for increased growth. However, he also asserts that the major setback in the French economy occurred in the agricultural sector, because of the importation of cheap food, and that growth in the industrial sector slowed less sharply than in agriculture.[53] It is impossible to disaggregate the overall effect of free trade on industrial growth into the components of growth in large-scale and small-scale production, respectively. Though large-scale, mechanized industry did develop during the second half of the nineteenth century, small-scale, handicraft manufacturing remained substantial and declined only slowly through this period.[54] Political economists, in their somewhat nostalgic advocacy of small-scale, quality manufactures, underestimated the role that large factories were about to play in the overall growth

---

Hachette, 1839) 2:50–63. But neither he nor his colleagues were as explicit about promoting specialized industrialization before the exhibition as they were in 1851.

51. Michael Stephen Smith, *Tariff Reform in France, 1860–1900: The Politics of Economic Interest* (Ithaca, N.Y.: Cornell University Press, 1980), 28–34.

52. Dunham, *Anglo-French Treaty;* Barrie M. Ratcliffe, "Napoleon III and the Anglo-French Commercial Treaty of 1860: A Reconsideration," *Journal of European Economic History* 2 (1973): 582–613.

53. Paul Bairoch, *Commerce extérieur et développement économique de l'Europe au XIXe siècle* (Paris: Mouton, 1976), 230–35.

54. Markovitch, "Revenu."

of the French economy. They were right, however, in seeing the production of luxury goods as an important and enduring component of that economy. Indeed, even when the government of the Third Republic revoked free trade in 1892, it also supported a revival of hand manufacturing through the art nouveau movement and the exhibition of 1900.[55] Chevalier, Blanqui, and their cohorts were among many who recognized and articulated the political as well as economic merits of small-scale, hand, and handicraft production for nineteenth-century France.

55. Debora L. Silverman, *Art Nouveau in Fin-de-Siècle France: Politics, Psychology, and Style* (Berkeley: University of California Press, 1989).

Conclusion

Taste and Consumption in Industrial Development

Though the Crystal Palace exhibition closed officially on 15 October 1851, its influence endured long after the strains of Handel's Hallelujah Chorus had faded away. The building itself enjoyed a long and full life, much to the gratification of London residents. Popular sentiment favored maintaining the Crystal Palace in Hyde Park as a permanent sanatorium for invalids, an art gallery, or a winter garden. But during the early planning stages the Exhibition Commission had promised to remove the building from its central London setting, and in 1852 this was duly carried out. The glass and iron structure was rebuilt in Sydenham on the southern outskirts of London, where it housed a variety of displays and events until fire consumed it in 1936.[1]

The importance of the Crystal Palace exhibition can be measured in the many international exhibitions of industry and world's fairs that followed it, right up to the present. The exhibition set a precedent for periodic, lavish displays of nations' industrial accomplishments and technological advances; its successors increasingly became palaces of consumption and places of public entertainment, as well as vehicles for government propaganda.[2] The Crystal Palace exhibition

---

1. John Tallis, *Tallis's History and Description of the Crystal Palace* (London and New York: John Tallis and Co., [1852]) 3:88–101; Christopher Hobhouse, *1851 and the Crystal Palace* (New York: E. P. Dutton and Co., 1937), 147–63.

2. Debora L. Silverman, *Art Nouveau in Fin-de-Siècle France: Politics, Psychology, and Style* (Berkeley: University of California Press, 1989); Patricia Mainardi, *Art and Politics of the Second Empire: The Universal Expositions of 1855 and 1867* (New Haven: Yale University Press, 1987); Miriam R. Levin, *Republican Art and Ideology in Late Nineteenth-Century France* (Ann Arbor: UMI Research Press, 1986); Rosalind H. Williams, *Dream Worlds: Mass Consumption in Late Nineteenth-*

has become a lasting symbol of modern industrialization, with both positive and negative connotations. That it continues to be interpreted and reinterpreted testifies to the rich implications of the exhibition for society, culture, and politics, as well as economics.

In France an immediate effect of the Crystal Palace exhibition was to intensify the competitive attitude surrounding French sponsorship of the next international exhibition of industry, in 1855. Louis-Napoleon Bonaparte was determined to outdo the English in the size and magnificence of the Paris exhibition, and he may be judged to have succeeded. In the longer term, the Crystal Palace exhibition established France's identity as an industrial power in an unprecedented manner. French people had always prided themselves on their good taste and the beauty of their manufactured goods; and the exhibition confirmed and strengthened this pride. It also suggested to many that France must exploit the art and taste of its industry in order to compete in world markets. With the Crystal Palace exhibition France found its market niche in the manufacture of luxury and consumer goods, to the extent that only nine years later the government overturned a centuries-long tradition of restrictions on international commerce and signed a free-trade agreement with Britain, France's most feared industrial rival. The merits of this policy are open to question, but the fact remains that the French economy handily withstood the increased foreign competition, and the industrial sector may indeed have benefited from it.

A less obvious, or at least less examined, consequence of the exhibition was explicit consideration in France for the role of the consumer, deriving from the acknowledged superiority of French products in their beauty of design and execution. Foreign as well as French writers attributed the tastefulness of French goods to the discriminating standards of French consumers. French industry produced beautiful things because French consumers settled for nothing less, the argument ran. French commentators liked to believe that they, and especially French women, were inherently more tasteful,

---

Century France (Berkeley: University of California Press, 1982); Richard D. Altick, *The Shows of London* (Cambridge, Mass.: Belknap Press, 1978); Werner Plum, *Les Expositions universelles au 19e siècle, spectacles du changement socio-culturel,* trans. Pierre Gallissaires (Bonn and Godesberg: Freidrich-Ebert-Stiftung, 1977); Jean Duvignaud, *Fêtes et civilisations* (Geneva: Weber, 1973).

more sensitive to beauty, than other nationalities. Assertions about inherent national characteristics hardly bear scrutiny, but an examination of the social and cultural factors behind French consumption reveals the significant influence of bourgeois women as well as men on industrial development.

This study has attempted to show that consumers in nineteenth-century France—essentially the bourgeoisie—favored the old styles associated with aristocratic and royal dominance, like Renaissance, Louis XIV, and Louis XV. They bought articles of furnishing and clothing that looked rich, because of their extensive ornamentation, artistic qualities, or expensive raw materials. Women in particular exercised these standards of consumption in their capacity as home-makers, the designated creators of pleasing and comfortable domestic interiors that also conveyed family status as bourgeois, as opposed to proletarian or aristocratic. For though inspired by the aristocracy past and present, bourgeois tastes were not entirely imitative. The emphasis on domestic comfort, rather than ostentatious display, distinguished bourgeois consumption from the aristocratic model. In addition, new technology both enhanced the comfort of home furnishings and lowered the costs of luxury and art objects, allowing bourgeois consumers to pursue goals of good taste and elegance within their comparatively limited incomes. In consumer goods industries, manufacturers' success depended as much on satisfying demand for beautiful and stylish products as on lowering costs through mass production methods. Thus men and women of the bourgeoisie indirectly encouraged hand and parcelized manufacturing. This tendency was further buttressed by theories and policies regarding French industry and economic development, theories that were inspired by the performance of France at the Crystal Palace exhibition.

The topics of taste and consumption do not lend themselves to the same kind of precision and certainty as production-related subjects. But they are no less important or revealing for that reason. To examine industrialization, as many scholars have, with almost no consideration for consumer demand is bound to result in an incomplete representation. Moreover, such efforts incorrectly tend to exclude women from any significant or influential role in industrialization. Analyzing consumption, especially in connection with production, discloses a new array of concerns and motivations for manufacturers, and it reinstates women in a history in which they were indeed active participants.

The study of consumption, then, serves as a key to understanding social groups—in this case the French bourgeoisie—and their social, economic, gender, and political relations. The potential for future research in this direction is great, if for no other reason than that comparatively little has been done to date. A good test of the main argument in this book—that the tastes of French consumers promoted hand manufacturing in French industry—would be to apply the same theory to England. That is, what were the demographics and standards of consumption in England, and did they promote factory production? A full treatment of this subject awaits further research and analysis, but existing secondary sources suggest some answers.

A few historians have already addressed the logic of a consumer revolution accompanying or even leading the English industrial revolution of the late eighteenth and early nineteenth centuries. The works of Elizabeth Waterman Gilboy and Neil McKendrick in particular argue that domestic demand for cheap articles of furnishing and clothing was the driving force behind the development of factories and mass production in England. A growing urban working class, and especially income-earning women and children, stimulated manufacturers to innovate in such a way as to churn out low-cost, standardized goods that these numerous consumers could afford to buy.[3] The demographics of English consumption thus differed markedly from those of France, where the population consisted predominantly of landed and nearly self-sufficient peasantry until the early twentieth century.[4] This difference also helps explain the different influences of labor supply on methods of production in England and France.

3. Elizabeth Waterman Gilboy, "Demand as a Factor in the Industrial Revolution," in *Facts and Factors in Economic History: Articles by Former Students of Edwin Francis Gay* (Cambridge: Harvard University Press, 1932); Neil McKendrick, "Home Demand and Economic Growth: A New View of the Role of Women and Children in the Industrial Revolution," in *Historical Perspectives: Studies in English Thought and Society,* ed. Neil McKendrick (London: Europa, 1974), 152–210.

4. Patrick O'Brien and Caglar Keyder, *Economic Growth in Britain and France, 1780–1914: Two Paths to the Twentieth Century* (London: George Allen and Unwin, 1978).

In England more of the working class was dependent solely upon manufacturing for survival, thus permitting the spread of mechanized production whereby workers had to adapt to constant machine attendance. By contrast, French workers frequently divided their labor time between agricultural and manufacturing pursuits. From a manufacturer's perspective, such an irregular and/or seasonal labor force, combined with consumer demand for beautiful and fashionable goods, rendered mass production—with its high start-up and overhead costs, its requirement of consistent and round-the-clock output, and its inability to adjust quickly to changes in demand—unattractive. A potent combination of factors, including consumer demand, labor supply, and working-class culture, influenced managerial decisions regarding production on a regional as well as national basis.[5]

But what about the English bourgeoisie? Though they may have been a minority among English consumers, did they use the same criteria in choosing household furnishings and clothing as their French counterparts? Some recent works of English social and economic history suggest that they did not; that the emphasis among bourgeois consumers in England was different from that in France, at least in the first half of the nineteenth century. For one thing, the function of bourgeois women as arbiters of taste and bearers of culture may have been less important in England than their role as moral exemplar. Of course, cultural activities in the home—like piano playing, needlework, drawing, and painting—were perfectly acceptable, even encouraged, for English females. But at least in provincial cities of the nineteenth century, cultural activities increasingly became part of the public and therefore male realm, particularly in the form of learned and literary societies. "Women's access to cultural production continued to be extremely limited," writes Janet Wolff in an essay on the separate spheres of culture in nineteenth-

5. Ibid.; Alain Cottereau, "The Distinctiveness of Working-Class Cultures in France, 1848–1900," in *Working-Class Formation: Nineteenth-Century Patterns in Western Europe and the United States,* ed. Ira Katznelson and Aristide R. Zolberg (Princeton: Princeton University Press, 1986), 111–54; Whitney Walton, "Working Women, Gender, and Industrialization: The Case of Lorraine Embroidering in the Nineteenth Century," *Journal of Women's History* 2 (Fall 1990): 42–65; George J. Sheridan, "Household and Craft in an Industrializing Economy: The Case of Silk Weavers of Lyons," in *Consciousness and Class Experience in Nineteenth-Century Europe,* ed. John M. Merriman (New York: Holmes and Meier Publishers, 1979), 107–28.

century England.[6] It is possible, then, that a slightly different gendering of cultural roles rendered tasteful consumption—perhaps consumption itself—a lesser priority for bourgeois women in England than in France.[7] This suggestion is consistent also with the influence of Nonconformist and Evangelical religion on the values and habits of a significant proportion of the English middle class.

Nonconformists were notably successful in commerce, industry, and finance, in part because they were excluded from other professions in Britain. In addition, they succeeded in enterprise because they valued sobriety, thrift, prudence, and hard work. These values translated into a domestic life that emphasized simplicity and comfort above style and elegance. Nonconformist men and women associated rich and opulent furnishings with a decadent, disreputable Anglican aristocracy. Their own preference for subdued clothing and comfortable furnishings reflected strong religious conviction and a deliberate distancing from their self-proclaimed social betters.[8] Though even religious families adopted more lavish furnishings with increasing wealth, and later generations gained entrance into larger social circles, many members of the British middle class shared the moral values and sense of class identity of the Nonconformists.

To characterize bourgeois consumption in nineteenth-century Britain as uniformly puritanical and insensitive to art or luxury would, of course, be misguided. Moreover, not all manufacturing in Britain was done in factories or by machines even after 1851. My aim has been to broaden conceptualizations of the bourgeoisie and of industrialization to include consumption. This was the lesson of the Crystal Palace exhibition for France, and one worth remembering today as consumption assumes even greater importance in the lives of increasing numbers of men, women, and children.

6. Janet Wolff, "The Culture of Separate Spheres: The Role of Culture in Nineteenth-Century Public and Private Life," in *The Culture of Capital: Art, Power and the Nineteenth-Century Middle Class,* ed. Janet Wolff and John Seed (Manchester: Manchester University Press, 1988), 127.

7. Patricia Branca, *Silent Sisterhood: Middle-Class Women in the Victorian Home* (Pittsburgh: Carnegie-Mellon University Press, 1975).

8. Leonore Davidoff and Catherine Hall, *Family Fortunes: Men and Women of the English Middle Class, 1780–1850* (Chicago: University of Chicago Press, 1987); Peter Mathias, *The First Industrial Nation: An Economic History of Britain, 1700–1914,* 2d ed. (London and New York: Methuen, 1983), 142–47.

Selected Bibliography

PRIMARY SOURCES

Archival Documents

Archives Nationales, Paris. Minutier Central.
VI 1230 (23 February 1870).
XII 1224 (5 January 1870).
XXXI 601 (4 March 1830).
XXXV 1098 (1 October 1829).
XXXV 1100 (24 April 1830).
XLI 1131 (8 April 1870, 27 April 1870).
XLI 1134 (4 July 1870, 9 July 1870).
XLV 954 (9 February 1870).
XLV 958 (24 November 1870).
LVI 648 (19 August 1828).
LVI 652 (6 March 1829).
LVI 657 (11 September 1829).
LVI 663 (15 June 1830).
LVI 664 (31 August 1830).
LVI 666 (6 November 1830, 17 December 1830).
LX 910 (1 February 1870).
LX 911 (14 March 1870, 30 March 1870).
LXXV 1338 (30 April 1870).
LXXV 1340 (29 June 1870).
LXXVI 946 (13 December 1869).
LXXVI 950 (10 August 1870).
LXXVII 587 (3 February 1830).
LXXVII 588 (7 April 1830).
LXXVII 590 (22 December 1830).

LXXIX 600 (18 February 1829).

LXXIX 607 (24 February 1830).

LXXIX 609 (4 May 1830).

CVII 963 (20 January 1870).

————. Fonds privés.

39 AP 4, Papiers Jullien de Paris, 1785–1863.

246 AP 40, Papiers de Mme H. Fortoul, 1849–1856.

AB XIX 3503, Comptes de ménagères, 1878–1900.

   K–M 106, Comptes de Mme Parizet à Taverny, 1884–1886.

   S 106, Lot de factures anciennes, 1878–1900.

————. Ministère de l'Agriculture et du Commerce.

F^{12} 2265, Orfèvrerie, bijouterie, bronzes, fabriques de plaqué d'or et d'argent, an III–1848, 1807–1821.

F^{12} 2282, Bronzes, ébenisterie, bijouterie, an IV–1848.

F^{12} 2334, Institutions, sociétés, dons, et prix en faveur des arts et manufactures, 1830–1870.

F^{12} 2410, Ébenisterie, papiers peints, an III–1852.

Bibliothèque de la Chambre de Commerce et d'Industrie de Paris. Chambre de Commerce et d'Industrie de Paris. Correspondance, 9 April 1846 to 18 June 1850.

Periodicals

L'Atelier

Le Bien-être universel

Le Conseiller des James

Le Conseiller des dames et demoiselles

Le Correspondant

Le Foyer domestique

La Gazette des femmes

La Gazette des salons

L'Illustration

Journal des économistes

Journal des mères et enfants

Les Modes parisiennes

Le Moniteur de la mode

Le Moniteur industriel

Le Musée des familles

Le National

Petit Courrier des dames

Le Petit Messager des modes
La Semaine

Books

Alq, Louise d'. *Le Savoir-vivre dans toutes les circonstances de la vie.* 2 vols. Paris: Bureaux des causeries familières, 1883.

The Art Journal. *The Art Journal Illustrated Catalogue: The Industry of All Nations, 1851.* London: J. Virtue, 1851. Reprinted as *The Crystal Palace Exhibition Illustrated Catalogue.* New York: Dover, 1970.

Audiganne, Armand. *L'Industrie contemporaine, ses caractères et ses progrès chez les différents peuples du monde.* Paris: Capelle, 1856.

———. *La Lutte industrielle des peuples.* Paris: Capelle, 1868.

Balzac, Honoré de. *Oeuvres complètes de M. de Balzac.* 26 vols. Paris: Les Bibliophiles de l'originale, 1965–76.

Blanqui, [Jérôme-]Adolphe. *Cours d'économie industrielle, 1838–1839.* Paris: Hachette, 1839.

———. *De la concurrence et du principe d'association.* Paris: Panckoucke, 1846.

———. *Des classes ouvrières en France pendant l'année 1848.* Paris: Firmin Didot frères, 1849.

———. *Histoire de l'Exposition des produits de l'industrie française en 1827.* Paris: Renard, 1827.

———. *Lettres sur l'Exposition universelle de Londres.* Paris: Cappelle, 1851.

[Brame, Caroline.] *Le Journal intime de Caroline B,* edited by Michelle Perrot and Georges Ribeill. Paris: Montalba, 1985.

Breton, Geneviève. *Journal, 1867–1871.* Paris: Ramsay, 1985.

Brisset, J.-A. "La Ménagère parisienne." In *Les Français peints par eux-mêmes.* Vol. 12. Paris: J. Philippart, n.d.

Celnart, Mme [Elisabeth Félicie Bayle-Mouillard]. *Manuel complet de la bonne compagnie.* Paris: Roret, 1834.

Chambre de Commerce et d'Industrie [de Paris].

———. *Statistique de l'industrie à Paris de l'enquête faite par la Chambre de commerce pour l'année 1860.* Paris: La Chambre, 1864.

———. *Statistique de l'industrie à Paris, 1847–48.* Paris, 1851.

Chevalier, Michel. *Examen du système commercial connu sous le nom du système protecteur.* 2d ed. Paris: Guillaumin, 1852.

———. *L'Exposition universelle de Londres.* Paris: Mathias, 1851.

———. *Lettres sur l'organisation du travail.* Paris: Capelle, 1848.

Comité central des artistes et des artistes-industriels. *Placet et mémoires relatifs à la question des beaux-arts appliqués à l'industrie.* Paris: Mathias, 1852.

Commission française sur l'Industrie des Nations. *Exposition universelle de 1851: Travaux de la Commission française sur l'Industrie des Nations.* 8 vols. Paris: Imprimerie impériale, 1853–56.

Coquelin, Charles, and Guillaumin, eds. *Dictionnaire de l'économie politique.* 2 vols. Paris: Guillaumin, 1852–53.

Dickinson's Comprehensive Pictures of the Great Exhibition of 1851. 2 vols. London: Dickinson Bros., 1854.

Encyclopédie méthodique par ordre de matières. Arts et métiers mécaniques. Vol. 4, *Planches du dictionnaire des arts et métiers.* Paris: Panckoucke, 1788.

Exposition des produits de l'industrie française en 1839. *Rapport du jury central.* 3 vols. Paris: Bouchard-Huzard, 1839.

Faucon, Emma. *Voyage d'une jeune fille autour de sa chambre: Nouvelle morale et instructive.* Paris: Maillet, 1860.

Flaubert, Gustave. *L'Éducation sentimentale.* 1867. Vol. 11 of *Les Oeuvres de Gustave Flaubert.* Lausanne: Éditions Rencontre, 1965.

———. *Madame Bovary.* 1857. Translated by Eleanor Marx-Aveling. New York: Jonathan Cape and Harrison Smith, 1930.

Garnier, Joseph, ed. *Le Droit au travail à l'Assemblée Nationale.* Paris: Guillaumin, 1848.

Garnier-Audiger. *Manuel du tapissier, décorateur, et marchand de meubles.* Paris: Roret, 1830.

Guillaumin, ed. *Dictionnaire du commerce et des marchandises.* 2 vols. Paris: Guillaumin, 1839–41.

Janet, Paul. *La Famille.* Paris: Michel Lévy, 1873.

Laborde, Léon de. *La Renaissance des arts à la cour de France: Études sur le seizième siècle.* 1850–55. 2 vols. New York: Burt Franklin, 1965.

———. *De l'union des arts et de l'industrie.* 2 vols. Paris: Imprimerie impériale, 1856.

Laboulaye, Charles. *Encyclopédie technologique: Dictionnaire des arts et manufactures de l'agriculture, des mines, etc.* 2 vols. 2d ed. Paris: Lacroix-Comon, [1853?].

———. *Essai sur l'art industriel.* Paris: Bureau du dictionnaire des arts et manufactures, 1856.

Labourieu, Théodore. *L'Organisation du travail artistique en France.* Paris: E. Dentu, 1863.

Lafarge, Mme [née Marie Capelle]. *Mémoires de Madame Lafarge.* Paris: Lévy, 1894.

Le Play, Frédéric. *Les Ouvriers européens.* Vol. 6. Paris: E. Dentu, 1878.

Ministère de l'Intérieur de l'Agriculture et du Commerce. *Annales du commerce extérieur: Faits commerciaux.* No. 20, *Exposition universelle de Londres en 1851.* Paris: Imprimerie impériale, 1853.

Musée industriel. *Description complète de l'Exposition générale des produits de l'industrie française faite en 1834.* 3 vols. Paris: Société polytechnique et du recueil industriel, 1836.

Pariset, Mme. *Manuel de la maîtresse de maison.* 3d ed. Paris: Audot, 1825.

———. *Nouveau Manuel complet de la maîtresse de maison.* Paris: Roret, 1852.

Rapports des délégués des ouvriers parisiens à l'Exposition de Londres en 1862. Paris: Chabaud, 1862–64.

Roubo, Alexandre [André Jacob?], fils. *L'Art du menuisier ébeniste.* Paris, 1774.

Sand, George. *My Life.* 1879. Translated by Dan Hofstadter. New York: Harper and Row, 1979.

Tallis, John. *Tallis's History and Description of the Crystal Palace.* 3 vols. London and New York: John Tallis and Co., [1852].

Véron, Eugène. *Histoire de l'Union centrale.* Paris: Debons, [1857?].

Zola, Émile. *La Curée.* 1871. Paris: Fasquelle, 1984.

SECONDARY SOURCES

Altick, Richard D. *The Shows of London.* Cambridge, Mass.: Belknap Press, 1978.

Appadurai, Arjun, ed. *The Social Life of Things: Commodities in Cultural Perspective.* New York: Cambridge University Press, 1986.

Bairoch, Paul. *Commerce extérieur et développement économique de l'Europe au XIXe siècle.* Paris: Mouton, 1976.

Benjamin, Walter. "Paris, Capital of the Nineteenth Century." In *Reflections: Essays, Aphorisms, Autobiographical Writings.* Translated by Edmund Jephcott. New York: Harcourt Brace Jovanovich, 1978.

———. "The Work of Art in the Age of Mechanical Reproduction." In *Illuminations.* Translated by Harry Zohn and edited by Hannah Arendt. New York: Harcourt, Brace and World, 1968.

Berg, Maxine, Pat Hudson, and Michael Sonenscher. *Manufacture in Town and Country before the Factory.* New York: Cambridge University Press, 1983.

Bergeron, Louis. "French Industrialization in the Nineteenth Century: An Attempt to Define a National Way." *Proceedings of the Annual Meeting of the Western Society for French History* 12 (1984): 154–63.

Boime, Albert. "Entrepreneurial Patronage in Nineteenth-Century France." In *Enterprise and Entrepreneurs in Nineteenth- and Twentieth-Century France,* edited by Edward C. Carter III, Robert Forster, and

Joseph N. Moody. Baltimore: Johns Hopkins University Press, 1976.

———. "The Teaching of Fine Arts and the Avant-Garde in France during the Second Half of the Nineteenth Century." *Arts Magazine* 60 (1985): 46–57.

———. "The Teaching Reforms of 1863 and the Origins of Modernism in France." *The Art Quarterly*, n.s., 1 (1977): 1–39.

Bouilhet, Henry. *L'Orfèvrerie française aux XVIIIe et XIXe siècles*. Paris: H. Laurens, 1910.

Bourdieu, Pierre. *La Distinction: Critique sociale du jugement*. Paris: Éditions de minuit, 1979.

Branca, Patricia. *Silent Sisterhood: Middle-Class Women in the Victorian Home*: Pittsburgh: Carnegie-Mellon University Press, 1975.

Breton, Yves. "Les Économistes, le pouvoir politique et l'ordre social en France entre 1830 et 1851." *Histoire, économie et société* 4 (1985): 233–52.

Cameron, Rondo E. "Economic Growth and Stagnation in France, 1815–1914." In *The Experience of Economic Growth: Case Studies in Economic History,* edited by Barry E. Supple. New York: Random House, 1963.

———. *France and the Economic Development of Europe, 1800–1914*. Princeton: Princeton University Press, 1961.

Clark, Timothy J. *The Absolute Bourgeois: Artists and Politics in France, 1848–1851*. London: Thames and Hudson, 1973.

Clouzot, Henri, and Charles Follot. *Histoire du papier peint en France*. Paris: C. Moreau, 1935.

Cottereau, Alain. "The Distinctiveness of Working-Class Cultures in France, 1848–1900." In *Working-Class Formation: Nineteenth-Century Patterns in Western Europe and the United States,* edited by Ira Katznelson and Aristide R. Zolberg. Princeton: Princeton University Press, 1986.

Darrow, Margaret H. "French Noblewomen and the New Domesticity, 1750–1850." *Feminist Studies* 5 (1979): 41–65.

Daumard, Adeline. *La Bourgeoisie parisienne de 1815 à 1848*. Paris: SEVPEN, 1963.

———. *Les Bourgeois et la bourgeoisie en France depuis 1815*. Paris: Aubier, 1987.

Davidoff, Leonore, and Catherine Hall. *Family Fortunes: Men and Women of the English Middle Class, 1780–1850*. Chicago: University of Chicago Press, 1987.

Demier, Francis. "Adolphe Blanqui, 1798–1854: Un Économiste libéral face à la Révolution industrielle." Ph.D. dissertation, University of

Paris, X–Nanterre, 1979.

Devaux, Yves. *L'Univers des bronzes et des fontes ornementales: Chef-d'oeuvres et curiosités, 1850–1920.* Paris: Pygmalion, 1978.

Dunham, Arthur Louis. *The Anglo-French Treaty of Commerce of 1860 and the Progress of the Industrial Revolution in France.* Ann Arbor: University of Michigan Press, 1930.

———. *The Industrial Revolution in France, 1815–1848.* New York: Exposition Press, 1955.

Faure, Alain. "Petit Atelier et modernisme économique: La Production en miettes au XIXe siècle." *Histoire, économie et société* 4 (1986): 531–57.

Fay, Charles Ryle. *Palace of Industry, 1851.* Cambridge: Cambridge University Press, 1951.

Forestier, Sylvie. "Art industriel et industrialisation de l'art: L'Exemple de la statuaire religieuse de Vendeuvre-sur-Barse." *Ethnologie française* n.s. 8, no. 2–3 (1978): 191–200.

Gaillard, Jeanne. *Paris, la ville, 1852–1870.* Paris: H. Champion, 1977.

Gay, Peter. *The Bourgeois Experience from Victoria to Freud.* 2 vols. New York: Oxford University Press, 1984–86.

Gerschenkron, Alexander. *Economic Backwardness in Historical Perspective.* Cambridge, Mass.: Belknap Press, 1962.

Grandjean, Serge. *L'Orfèvrerie du XIXe siècle en Europe.* Paris: Presses universitaires de France, 1962.

Hellerstein, Erna Olafson, Leslie Parker Hume, and Karen M. Offen, eds. *Victorian Women: A Documentary Account of Women's Lives in Nineteenth-Century England, France, and the United States.* Stanford: Stanford University Press, 1981.

Hobhouse, Christopher. *1851 and the Crystal Palace.* New York: E. P. Dutton and Co., 1937.

Kaplan, Steven Laurence, and Cynthia J. Koepp, eds. *Work in France: Representations, Meaning, Organization, and Practice.* Ithaca, N.Y.: Cornell University Press, 1986.

Kindleberger, Charles P. *Economic Growth in France and Britain, 1851–1950.* Cambridge: Harvard University Press, 1964.

Kjellberg, Pierre. *Les Bronzes du XIXe siècle: Dictionnaire des sculpteurs.* Paris: Les Editions de l'amateur, 1987.

Landes, David S. *The Unbound Prometheus.* London: Cambridge University Press, 1969.

Ledoux-Lebard, Denise. *Les Ébenistes parisiens du XIXe siècle, 1795–1870.* 2d ed. Paris: F. de Nobèle, 1965.

Levasseur, Émile. *Histoire des classes ouvrières et de l'industrie en France de 1789 à 1870.* 2 vols. Paris: A. Rousseau, 1904.

Levin, Miriam R. *Republican Art and Ideology in Late Nineteenth-Century France.* Ann Arbor: UMI Research Press, 1986.

Lévy-Leboyer, Maurice. "Les Processus d'industrialisation: Le Cas de l'Angleterre et de la France." *Revue historique* 239 (1968): 281–98.

McKendrick, Neil. "Home Demand and Economic Growth: A New View of the Role of Women and Children in the Industrial Revolution." In *Historical Perspectives: Studies in English Thought and Society,* edited by Neil McKendrick. London: Europa, 1974.

Magraw, Roger. *France, 1815–1914: The Bourgeois Century.* New York: Oxford University Press, 1986.

Mainardi, Patricia. *Art and Politics of the Second Empire: The Universal Expositions of 1855 and 1867.* New Haven: Yale University Press, 1987.

Markovitch, T. J. "Le Revenu industriel et artisanal sous la Monarchie de Juillet et le Second Empire." *Économies et sociétés,* série AF-8 (April 1967): 85–87.

Martin-Fugier, Anne. *La Bourgeoisie.* Paris: Grasset, 1983.

———. *La Place des bonnes.* Paris: Grasset, 1979.

Mayer, Arno. *The Persistence of the Old Regime: Europe to the Great War.* New York: Pantheon, 1981.

Merriman, John M., ed. *Consciousness and Class Experience in Nineteenth-Century Europe.* New York: Holmes and Meier Publishers, 1979.

Miller, Daniel. *Material Culture and Mass Consumption.* Oxford and New York: Basil Blackwell, 1987.

Miller, Michael B. *The Bon Marché: Bourgeois Culture and the Department Store, 1869–1920.* Princeton: Princeton University Press, 1981.

Morazé, Charles. *The Triumph of the Middle Classes.* Translated by George Wiedenfeld. Cleveland: World Publishing Co., 1966.

Nouvel, Odile. *Wallpapers of France, 1800–1850.* Translated by Margaret Timmers. New York: Rizzoli, 1981.

O'Brien, Patrick, and Caglar Keyder. *Economic Growth in Britain and France, 1780–1914: Two Paths to the Twentieth Century.* London: George Allen and Unwin, 1978.

Perrot, Marguerite. *Le Mode de vie des familles bourgeoises, 1873–1953.* 2d ed. Paris: Presses de la fondation nationale des sciences politiques, 1982.

Perrot, Michelle, Alain Corbin, Roger-Henri Guerrand, Catherine Hall, Lynn Hunt, and Anne Martin-Fugier. *De la Révolution à la Grande Guerre.* Vol. 4 of *Histoire de la vie privée,* edited by Philippe Ariès and Georges Duby. Paris: Éditions Seuil, 1987.

Perrot, Philippe. *Les Dessus et les dessous de la bourgeoisie.* Paris: Arthème Fayard, 1981.

Pevsner, Nikolaus. *Academies of Art Past and Present*. Cambridge: Cambridge University Press, 1940.

———. *High Victorian Design: A Study of the Exhibits of 1851*. London: Architectural Press, 1951.

Philadelphia Museum of Art. *The Second Empire, 1852–1870: Art in France under Napoleon III*. Philadelphia: Philadelphia Museum of Art, 1978.

Price, Roger. *The Economic Modernisation of France*. New York: Wiley, 1975.

Ratcliffe, Barrie M. "Napoleon III and the Anglo-French Commercial Treaty of 1860: A Reconsideration." *Journal of European Economic History* 2 (1973): 582–613.

Roche, Daniel. *The People of Paris*. Translated by Marie Evans with Gwynne Lewis. Berkeley: University of California Press, 1987.

Roehl, Richard. "French Industrialization: A Reconsideration." *Explorations in Economic History* 13 (1976): 233–81.

Sabel, Charles F., and Jonathan Zeitlin. "Historical Alternatives to Mass Production: Politics, Markets and Technology in Nineteenth-Century Industrialization," *Past and Present* 108 (1985): 133–76.

Samuel, Raphael. "Workshop of the World: Steam Power and Hand Technology in Mid-Victorian Britain." *History Workshop* 3 (Spring 1977): 6–72.

Silverman, Debora L. *Art Nouveau in Fin-de-Siècle France: Politics, Psychology, and Style*. Berkeley: University of California Press, 1989.

Simmel, Georg. *The Philosophy of Money*. 1904. Translated by Tom Bottomore and David Frisby. London and Boston: Routledge and Kegan Paul, 1978.

Smith, Bonnie G. *Ladies of the Leisure Class: The Bourgeoises of Northern France in the Nineteenth Century*. Princeton: Princeton University Press, 1981.

Smith, Michael Stephen. *Tariff Reform in France, 1860–1900: The Politics of Economic Interest*. Ithaca, N.Y.: Cornell University Press, 1980.

Sorlin, Pierre. *La Société française*, vol. 1, *1840–1914*. Paris: Arthaud, 1969.

Sparling, Tobin Andrews. *The Great Exhibition: A Question of Taste*. New Haven: Yale Center for British Art, 1982.

Tilly, Louise A., and Joan W. Scott. *Women, Work, and Family*. New York: Holt, Rinehart and Winston, 1978.

Vanier, Henriette. *La Mode et ses métiers: Frivolités et luttes des classes, 1830–1870*. Paris: Armand Colin, 1960.

Veblen, Thorstein. *The Theory of the Leisure Class: An Economic Study in the Evolution of Institutions*. New York: Macmillan Co., 1899.

Walton, Whitney. "Feminine Hospitality in the Bourgeois Home of Nineteenth-Century Paris," *Proceedings of the Western Society for French History* 14 (1987): 197–203.

———. "The Persistence of French Handicraft Industries and the Crystal Palace Exhibition of 1851." Ph.D. diss., University of Wisconsin-Madison, 1983.

———. "Political Economists and Specialized Industrialization during the French Second Republic, 1848–1852." *French History* 3, no. 3 (1988): 293–311.

———. " 'To Triumph before Feminine Taste': Bourgeois Women's Consumption and Hand Methods of Production in Mid-Nineteenth-Century Paris." *Business History Review* 60 (Winter 1986): 541–63.

Weissbach, Lee Shai. "Artisanal Responses to Artistic Decline: The Cabinetmakers of Paris in the Era of Industrialization." *Journal of Social History* 16 (Winter 1982): 67–81.

Williams, Rosalind H. *Dream Worlds: Mass Consumption in Late Nineteenth-Century France.* Berkeley: University of California Press, 1982.

Zeldin, Theodore. *France, 1848–1945.* 4 vols. Oxford: Oxford University Press, 1980.

Index

Compositor: A-R Editions, Inc.
Text: 10.5/13 Bembo
Display: Bembo
Printer: Braun-Brumfield, Inc.
Binder: Braun-Brumfield, Inc.